28th Micromechanics and Microsystems Europe Workshop (MME 2017)

AA006700

Journal of Physics: Conference Series Volume 922

Uppsala, Sweden
23 – 25 August 2017

ISBN: 978-1-5108-5270-9
ISSN: 1742-6588

Printed from e-media with permission by:

Curran Associates, Inc.
57 Morehouse Lane
Red Hook, NY 12571

CURRAN ASSOCIATES INC. proceedings.com

Some format issues inherent in the e-media version may also appear in this print version.

Printed by Curran Associates, Inc. (2017)

Institute of Physics
Dirac House, Temple Back
Bristol BS1 6BE UK

Phone: 44 1 17 929 7481
Fax: 44 1 17 920 0979

techtracking@iop.org

Additional copies of this publication are available from:

Curran Associates, Inc.
57 Morehouse Lane
Red Hook, NY 12571 USA
Phone: 845-758-0400
Fax: 845-758-2633
Email: curran@proceedings.com
Web: www.proceedings.com

Table of contents

Volume 922

28th Micromechanics and Microsystems Europe Workshop
23–25 August 2017, Uppsala, Sweden

Accepted papers received: 30 October 2017
Published online: 23 November 2017

This volume includes selected papers from MME 2017 – the 28th Micromechanics and Microsystems Europe Workshop. MME dates all the way back to 1989, when 25 participants gathered to hear about trends in the new and emerging field of microsystems technology in Twente, The Netherlands. In 2000, MME was held in Uppsala for the first time, and, seventeen years on from that gathering, we were delighted to welcome MME back to us.

Forty high quality submissions were accepted for the conference from authors affiliated to 13 countries and a selection of those are collected here. The works span all aspects of MEMS and microsystems technology, including design and fabrication, simulation and modelling, experimental evaluation, and commercial translation.

We acknowledge the important contributions made by our four invited speakers, who discussed exciting recent developments and their future visions for microsystems technology.

Our conference Gala Dinner was held at Norrlands nation, one of the typical old student clubs of Uppsala. With the strong investments in new facilities, Uppsala University is both the oldest and the newest University in Sweden, and ranked the highest of the comprehensive Universities of our nation. Being in the cradle of Sweden and an old University City, unique cultural treasures and an exciting history can be found in this city of knowledge and inspiration. Uppsala has retained its small-town charm while offering a big city's selection of shops, restaurants and other entertainment.

Uppsala University has received generous support from industrial partners and the Institute of Physics. We sincerely thank them for their assistance with the hosting of MME 2017.

Many thanks also to the Programme Committee, our Local Organizing Committee for their assistance with event logistics, and to the management and staff of Academic Conferences.

We hope that you will enjoy this Proceedings, and look forward to seeing you in Uppsala, Sweden, for MME 2017.

Klas Hjort and Maria Tenje
Local Co-Chairs – MME 2017

IOP Conf. Series: Journal of Physics: Conf. Series **922** (2017) 011002 doi:10.1088/1742-6596/922/1/011002

Peer review statement

All papers published in this volume of *Journal of Physics: Conference Series* have been peer reviewed through processes administered by the proceedings Editors. Reviews were conducted by expert referees to the professional and scientific standards expected of a proceedings journal published by IOP Publishing.

High pressure inertial focusing for separation and concentration of bacteria at high throughput

F.J. Cruz and K. Hjort

Engineering Sciences, Uppsala University, Ångström Laboratoriet, Uppsala, Sweden

javier.cruz@angstrom.uu.se; klas.hjort@angstrom.uu.se

Abstract. Inertial focusing is a phenomenon where particles migrate across streamlines in microchannels and focus at well-defined, size dependent equilibrium points of the cross section. It can be taken into advantage for focusing, separation and concentration of particles at high through-put and high efficiency. As particles decrease in size, smaller channels and higher pressures are needed. Hence, new designs are needed to decrease the pressure drop. In this work a novel design was adapted to focus and separate 1 µm from 3 µm spherical polystyrene particles. Also 0.5 µm spherical polystyrene particles were separated, although in a band instead of a single line. The ability to separate, concentrate and focus bacteria, its simplicity of use and high throughput make this technology a candidate for daily routines in laboratories and hospitals.

Keywords. Particle separation, Bacteria separation, Inertial focusing, Microfluidic channel, High pressure.

1. Introduction

Inertial focusing is a phenomenon where particles migrate across streamlines in microchannels and focus at well-defined, size dependent equilibrium points of the cross section. In a straight system it is caused by the balance of two forces [1], Fig. 1.

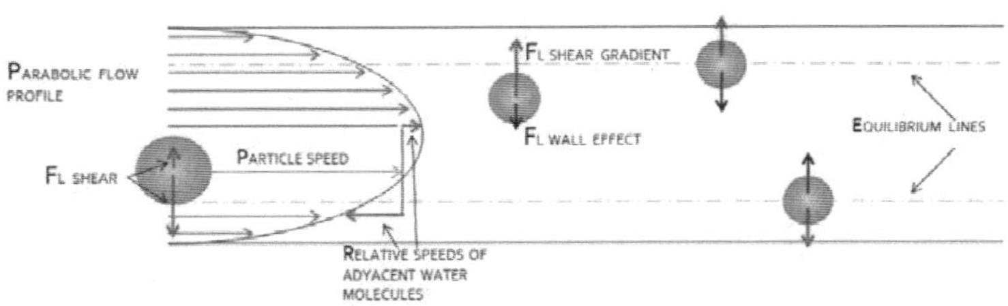

Figure 1. Main forces on a particle in a straight microchannel.

a shear lift force directed towards the walls of a channel due to the shape of the velocity profile ($F_{L\ SHEAR\ GRADIENT}$) and a wall lift force directed towards the center due to interactions of the streamlines with the wall ($F_{L\ WALL\ EFFECT}$). The net lift (F_L) force was predicted by Asmolov [2].

$$F_L = \frac{4\rho C_L {U_f}^2 {a_p}^4}{{D_h}^2} \qquad \text{Eq. 1}$$

where ρ is the fluid density, C_L is the lift coefficient which is a function of the particle position across the channel cross-section and the channel Reynolds number, U_f is the average flow velocity, a_p is the particle diameter and $D_h = \frac{4(hw)}{h+w}$ the hydraulic diameter of the channel, with h its height and w its width. In curved channels the Dean flow enhances the lateral motion of particles and reduces the focus length [3].

We previously showed a set of scaling factors (Table 1) that maintain the magnitude of the lift forces and allow the transformation of a system that works for a certain size of particles into a system that successfully focuses smaller sizes. We also supported the idea with experimental results showing alignment of 1 µm particles and *Escherichia coli (E. coli)* in a spiral microchannel. However, the high pressure was on the limit for the *E. coli*. Its viability was similar before and after being focused at 50 µl/min (70 bar). However, at 100 µl/min (150 bar) 90% of the bacteria died. [4]

Table 1. Scaling law to transform a design that works for certain particle size to target another size. [4]

Scaling Relations	Particle size	Height	Width	Flow rate	Focus length	Average speed	ΔP
Scale factor	X	X	X	X	X	X^{-1}	X^{-2}

In this paper, the aim is to show an alternative design that allows for inertial focusing faster and at lower pressure drops than with previous high throughput designs.

2. Experimental details

Microfluidic chips that could tolerate pressure up to 200 bar were dry etched (Tegal 110 S/DE) in silicon and anodically bonded to borosilicate glass, diced and finally silica capillaries were glued as connections.

The microchannel was 6.6 µm deep and consisted of a series of expansions-contractions along 3-4.6 mm. The narrow parts were 20 µm that expanded until 60 µm.

Fluorescent polystyrene particles with diameters of 3, 1 and 0.5 µm and *E. coli* carrying Yellow Fluorescent Protein were suspended in deionized water at a concentration of 10^6, 10^7, 10^8 and 10^8 particles/ml, respectively.

A high pressure HPLC pump (Waters, model 515) was used to pump the samples.

IOP Conf. Series: Journal of Physics: Conf. Series **922** (2017) 012001 doi:10.1088/1742-6596/922/1/012001

3. Results

A design of straight channel with varying width showed excellent results already at 50 µl/min and approx. 30 bar, Fig. 2.

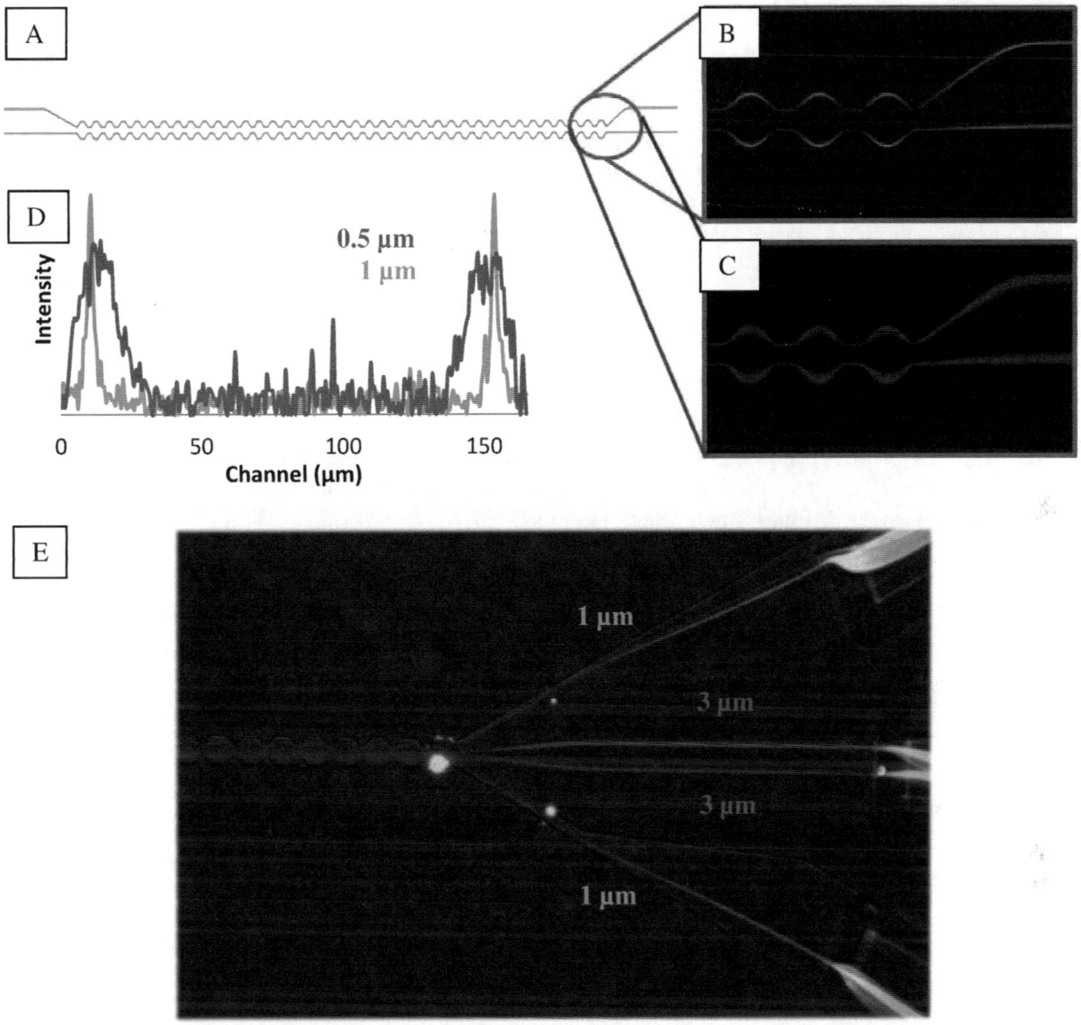

Figure 2. (A) General view of a straight channel (4.6 mm in length) with varying width (B) Performance with 1 µm particles at 50 µl/min. (C) Performance with 0.5 µm particles at 50 µl/min (D) Analysis of the intensity signal. (E) Separation of 1 (green) and 3 (red) µm particles at 150 µl/min and approx. 100 bar.

4. Discussion

The design with straight channels with varying width is still under evaluation, Fig. 2. A preliminary study showed not only a faster alignment of 1 µm particles than in curved models but it also required less pressure for the same flow rate. It could also concentrate 0.5 µm particles to a narrow band enabling their separation.

The design can be shortened relieving some pressure drop. The system is expected to focus 1 µm particles at 50 µl/min demanding around 20 bar.

The separation between 1 and 3 µm is quite large, allowing for further discrimination of intermediate sizes.

5. Conclusion

In this work we used straight microchannels that vary their width. The new design aligned 1 µm particles faster that the spiral model and required less pressure. Furthermore, although not completely aligned, 0.5 µm particles were concentrated in a band and were separated from the main stream.

The ability to separate, concentrate and focus bacteria, its simplicity of use and high throughput make this technology a candidate for daily routines in laboratories and hospitals.

Acknowledgements
The project has received funding from the European Union's Horizon 2020 research and innovation program under grant agreement no. 644669.

References

[1] H. Amini, W.H. Lee, and D. Di Carlo, Inertial microfluidic physics, Lab Chip 14 (2014) 2739.
[2] E.S. Asmolov, The inertial lift on a spherical particle in a plane Poiseuille flow at large channel Reynolds number, J. Fluid Mech., 381 (1999), 63–87.
[3] J.M. Martel and M. Toner, Inertial focusing dynamics in spiral microchannels, Phys. Fluids 24 (2012) 032001.
[4] F.J. Cruz, S. Hooshmand Zadeh, T. Graells, M. Andersson, J. Malmström, Z.G. Wu, and K. Hjort, High pressure inertial focusing for separating and concentrating bacteria at high throughput, J. Micromech. Microeng. (accepted manuscript April 2017, http://iopscience.iop.org/article/10.1088/1361-6439/aa6b14/pdf).

IOP Conf. Series: Journal of Physics: Conf. Series **922** (2017) 012002 doi:10.1088/1742-6596/922/1/012002

Design and Development of a Novel Knudsen Compressor as a Part of a Joule-Thomson Cryocooler

Indra Muthuvijayan, Mikael Antelius, Are Björneklett, Peter Nilsson and Robert Thorslund

APR Technologies, 745 39, Sweden

E-mail: indra@aprtec.com

Abstract. This paper presents the design and development of a novel Knudsen compressor, with no moving parts, as a part of a Joule-Thomson cryocooler. The compressor works by using the Knudsen diffusion principle and includes a combination of graphene-based layers and Knudsen membranes in a particular fashion to pressurize the fluid. The Knudsen membrane for this application was selected by testing several commercially available materials. Prototypes of single stage and a multistage compressors are presented together with experimental evaluations. Insights on a Tube-in-Tube heat exchanger, as another part a the Joule-Thomson cryocooler, intended to integrate with the Knudsen compressor, are also presented.

1. Introduction

Thermal management is a critical part of spacecraft design and performance. Generally, every piece of equipment has a specific temperature range which is appropriate for its working condition. A spacecraft contains components/systems that are required to be significantly cooled, down to cryo temperatures (below -150 °C). These components are often critical for the mission such as optical subsystem in an observatory spacecraft. For example, a telescope detector is required to be kept at cryogenic temperature to detect very faint IR waves from long distance [1]. A cryogenic cooler is a type of thermal management system that maintains cryogenic temperature at a desired location. Conventional cryogenic systems are generally large and heavy, mostly because of its compressor (a small and low mass compressor replacing the conventional one are always welcomed). A Knudsen compressor can be one such replacement for a Joule-Thomson cryocooler. The Knudsen compressor developed at APR Technologies is small, lightweight, reliable and consumes less power than the conventional cryo compressor [2]. Since this compressor does not involve any moving parts, it does not introduce vibrations or wear and tear, which reduces its potential risk of failure compared to compressors with moving parts.

2. Joule-Thomson Cryocooler

Cryogenics have shown tremendous progress in the last 25 years [2, 3]. A cryocooling cycle, in general, is compared to the reversed Carnot cycle because its highest efficiency involves isentropic (constant entropy) processes. It is a virtual cycle because, in actual cases, the isentropic (ideal) processes are not possible as the entropy of the universe (*i.e.* system + surroundings) always increases. In the basic principle behind the reversed Carnot cycle, the heat is absorbed from the cold storage and transferred to the hot by applying work into the system. It is the exact opposite of the fundamental heat engine [2, 3, 4, 5]. The schematic of a Joule-Thomson (JT) cryocooler [2] is analogous to an actual refrigeration system consisting of a compressor, a heat exchanger, an expansion valve and a reservoir. The difference between a JT cryocooler with that of other refrigeration system is the throttling/expansion process that takes place in the JT valve. This expansion process is maintained at constant enthalpy unlike in other expansion valves. The cold output of the JT cryocooler is achieved by the fluid after going through several phases namely, compression, pre-cooling, heat exchange and expansion. The cumulative performance of each of the components in the system adds up to determine the efficiency of the entire cryocooler system.

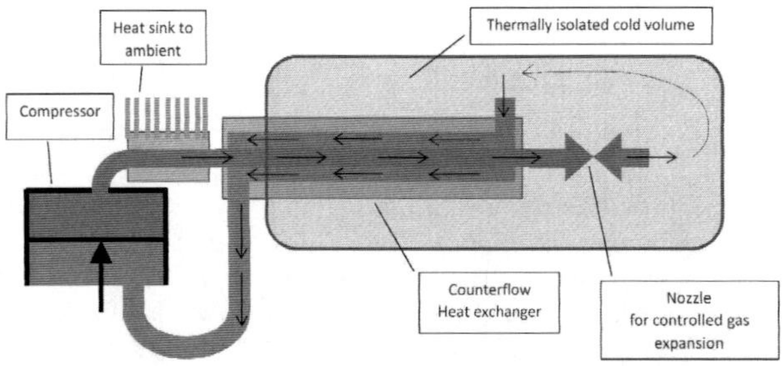

Figure 1. Schematic of a Joule Thomson cryocooler. Blue colour indicates the cold flow and the green indicates the hot flow. The cryo device is kept inside the thermally isolated volume, at the coldest end of the nozzle.

The Joule-Thomson cooling cycle is based on the principle of expanding a pressurized gas through a special valve called the Joule-Thomson expansion valve, where throttling or fluid expansion takes place. Figure 1 represents a simple schematic of a JT cooler. The incoming gas is compressed while maintaining the same inlet and outlet temperatures. The pressurized gas is fed into series of components namely, pre-cooler to bring down the gas to room temperature, heat exchanger where the gas further cooled down and finally to the JT valve where the cooled pressurized gas is expanded maintaining the enthalpy constant (JT effect). The cycle is complete when the gas collects the heat from the targeted device. The JT effect can be seen only in non-ideal gases. All the known gases are non-ideal, which implies the choice of the working fluid for the cryocooler is quite wide. However, the choice is narrowed by the inversion temperature property of the fluid which determines the energy required for it to achieve a positive JT effect (cooling effect) and cooling power.

3. Knudsen Compressor Overview

A Knudsen compressor compresses the gas using Knudsen diffusion. Knudsen compression takes place when a temperature gradient is imposed across a porous and low thermal conductivity transpiration membrane. The Knudsen membrane is selected based on the operating pressure and the mean free molecular path of the working fluid. To have a Knudsen diffusion, the pore size in the Knudsen membrane should be comparable (not larger than 10 times [6]) to the mean free molecular path of the gas flowing through it. Knudsen diffusion occurs when the molecules collides with the walls of the membrane rather than colliding with each other. The movement of molecules from the region with lower temperature to that of the higher temperature due to Knudsen diffusion is called thermal transpiration. The molecules get warmer as they reach the hot end and this is coordinated with the pressure increase that is proportional to the square root of the temperature. The governing equation of this flow is given as $P_H/P_C = (T_H/T_C)^{1/2}$. A higher pressure at the inlet of the JT valve leads directly to a lower temperature at the valve outlet [6]. Therefore, the compressor indirectly determines the capability of the cooling system.

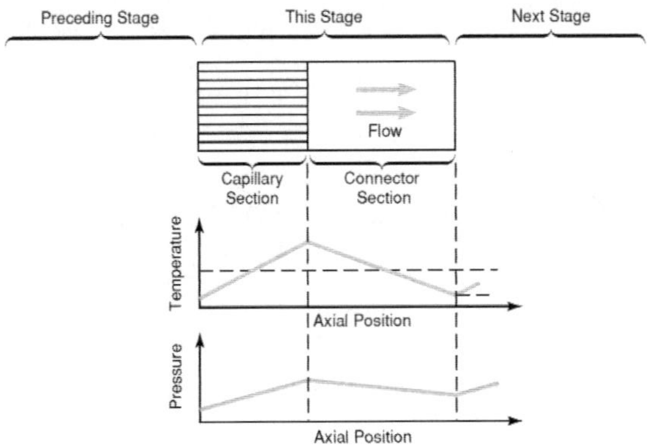

Figure 2. Temperature and pressure profile across a single stage of a compressor. Capillary and connector section refers to the Knudsen and insulation membrane respectively.

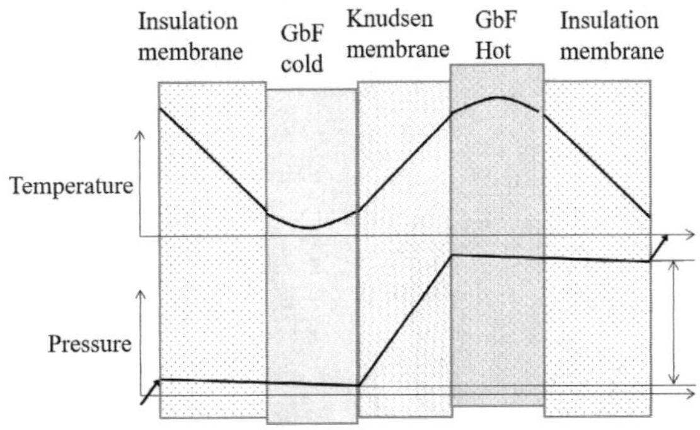

Figure 3. Temperature and pressure profile in a multistage compressor

The pressure inside a Knudsen membrane is built up due to the flow restriction of the gas molecules. Variation of the temperature, from low to high, across the membrane facilitates the thermal transpiration and the rise in temperature is reflected as a pressure rise as shown in Figure 2 [6, 7]. The Knudsen compressors that APR Technologies has designed consists of one Knudsen membrane per stage, surrounded by hot/cold Graphene-based Films(GbF) and thermally insulated to the next stage by membranes, as shown in Figure 3. The Knudsen membrane is cooled and heated with the help of GbF at its leading and trailing surfaces respectively. Copper rails (not shown) support these layers mechanically and is used as thermal busses to transport heat in and out of the GbF layers.

4. Design Procedure

A study carried out previously at APR technologies derived a targeted requirement for the JT cryocooler. This serves as an input to the design of the system level parts. The compressor should have an output of 10 bar for the correct function of other components of the JT-cryocooler as mentioned below. The cooling power of the cryocooler is required to dissipate 40 mW and maintain the cryo temperature below 150 K [7].

- Gas mixture: Nitrogen / hydrocarbon / (possibly noble gas)
- Pressure ratio: 1:10
- Heat load: 40 mW
- Heat exchanger efficiency: 95%
- Gas flow rate at normal conditions: 100-200 ml/min

The following compressor design and results will help to estimate the number of stages required to satisfy the compressor output pressure requirements.

4.1. Single-stage Compressor

A single stage compressor was designed as a technology demonstrator and for selecting the right Knudsen membrane. A schematic diagram of the single stage Knudsen compressor is shown in Figure 4, where the single Knudsen membrane is fitted into an Aluminium casing. Heat is applied with the help of Kanthal wire at one end and the other end is cooled with cold-water current inside copper tubes. The working gas used for this compressor is N_2 which has a positive JT coefficient at normal temperatures.

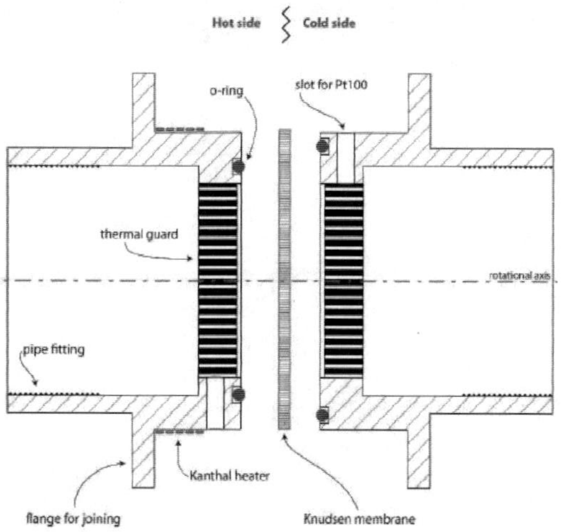

Figure 4. Various parts of a single stage Knudsen compressor

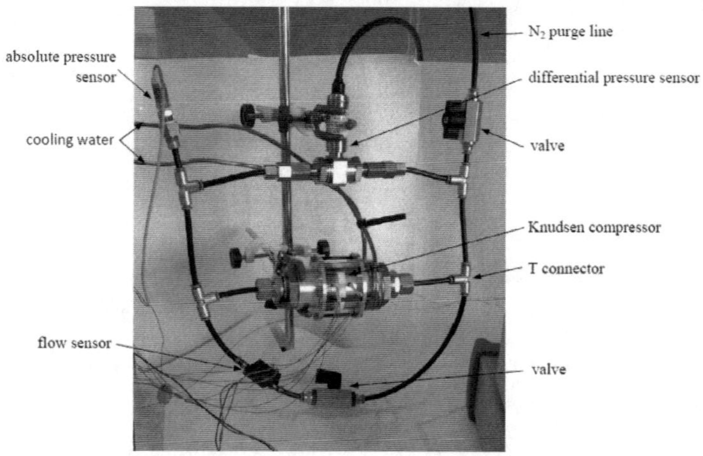

Figure 5. Test setup for a single stage Knudsen compressor

A suitable sealant was employed at vulnerable points to avoid leakage. The test setup of the single stage Knudsen compressor with the sensing elements can be seen in Figure 5. The hoses are connected to the gaseous source at the end. Flow control valves are equipped in the setup for letting in the gas and to test the compressor.

4.2. Multi-stage Compressor

Multiple Knudsen stages are required to realize a high-pressure ratio of the compressor. As the working fluid passes the Knudsen stages, the pressure is increased in tandem with the temperature. This implies that the temperature of the media needs to be lowered between the stages in order to have continuous flow. The design of multi-stage compressor is therefore different from the single stage compressor described above. The changes include the lighter plastic casing, the rectangular shape of the Knudsen membrane and the improved heating and cooling of the gas using porous sheets of graphene based films attached to copper heating/cooling blocks. These changes are made to easily increase the number of stages as required. As indicated in Figure 6, the stages are separated by a free flow and thermally insulating membrane (A). The stages themsels consist of a cooling GbF on the leading face (B), a Knudsen diffusion membrane (C) and a heating GbF on the trailing face (D). The multistage design uses a perforated GbF to facilitate loss-free gas flow. This is distinguished from the previously described single stage design, where a circumferential Aluminium contact serves as a thermal contact. To share a common cold and hot region, the GbF and other layers are bonded into a copper slabs. A multistage design involves stacking of layers in the order B-C-D[-A-B-C-D]$_n$

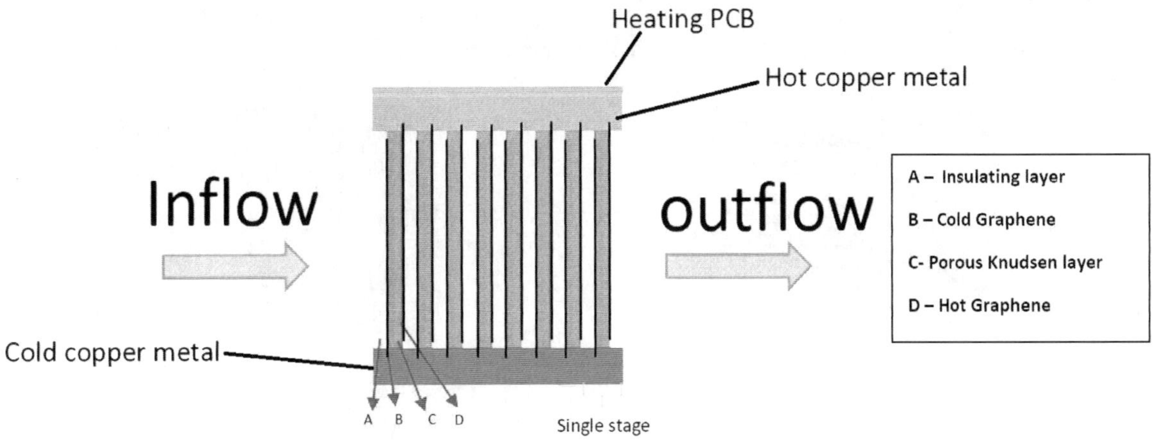

Figure 6. Multi-stage Knudsen compressor design.

Figure 7. CAD model of the multistage compressor casing.

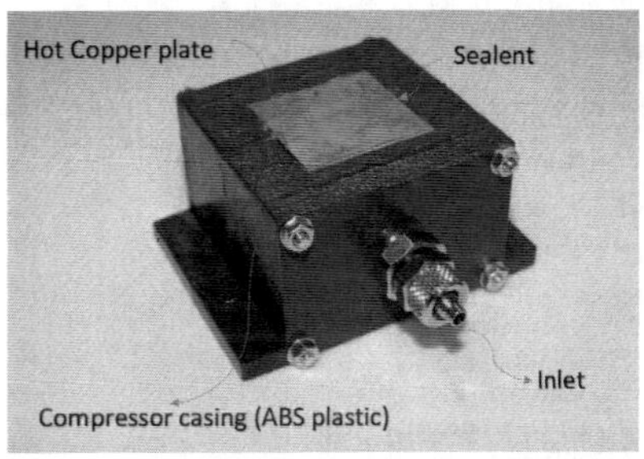

Figure 8. Assembled multistage compressor.

A four-stage and a three-stage Knudsen compressors were designed and fabricated. Figure 6 shows an eight stage cascade Knudsen compressor. The variation of pressure and temperature across the layers of the multistage Knudsen compressor is illustrated in Figure 3. At the end of every stage, there is a pressure rise and slight pressure drop before it enters the next stage. This drop-in pressure is due to the conversion of pressure energy into flow energy. The harmonic temperature variation across the layers corresponds to the compressor pressure accumulation.

4.3. Compressor Assembly and Test Setup

A viable casing for a four-stage compressor with proper support to hold everything together at the right position was 3D printed. The material used for 3D printing was ABS plastic. The CAD model of the final casing design is shown in Figure 7. The compressor core layers are bonded to the copper with the help of thermally conductive adhesive. It is then bonded to the compressor case to make the compressor leak proof. Copper on the top and bottom surface of the compressor are used for thermal contact for heating and cooling. The assembled compressor can be seen in Figure 8. The test setup for the compressor includes a cooling plate connected to the bottom of the compressor, a heater connected at its top surface, hoses connected to the source gas, sensing electronics connected to a computer and power supplies. Figure 9 shows the

IOP Conf. Series: Journal of Physics: Conf. Series **922** (2017) 012002 doi:10.1088/1742-6596/922/1/012002

compressor mounted between the cooling plate and PCB heater. The laboratory test setup of the compressor, the sensing elements and data acquisition system can be seen in Figure 10.

Figure 9. Compressor mounted on a cooling plate.

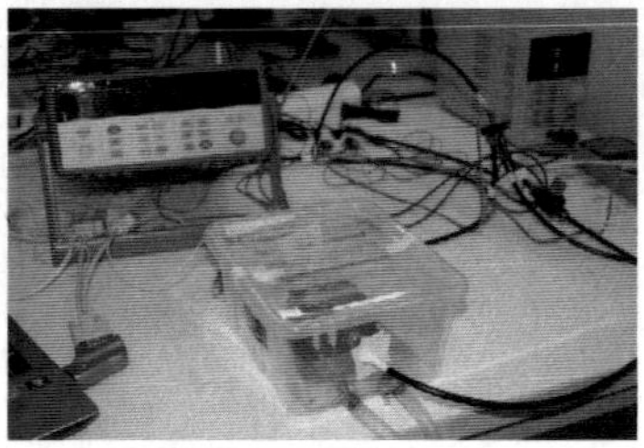

Figure 10. Laboratory test setup of a multistage compressor.

5. Compressor Testing

5.1. Thermal Testing

To achieve thermal transpiration, the Knudsen membrane is cooled and heated at its leading and trailing surfaces respectively. The temperature profile of the compressor core is simulated in a finite element tool (NASTRAN/PATRAN). Figure 11 shows the temperature profile of the simulated model using the boundary conditions of 353 K at the hot GbF and 273 K at the cold GbF. For reference, the measured core temperatures were 76°C (349 K) and 10°C (283 K) in hot and cold end, respectively.

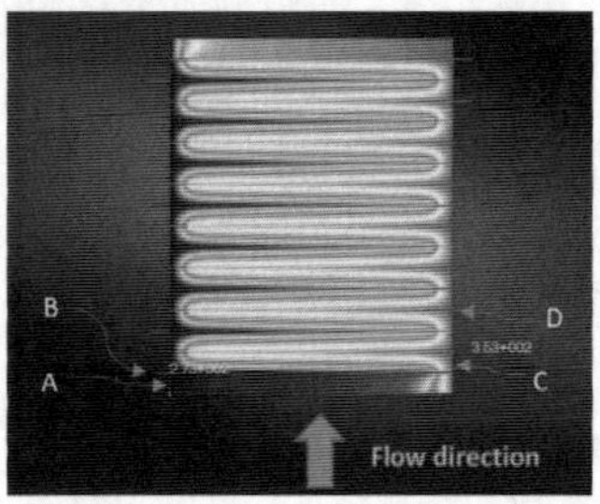

Figure 11. Multi stage thermal simulation: Temperature

Figure 12. Multi stage thermal observation using IR camera.

The temperature measurement by the IR camera is dependent on the emissivity of the observed materials that is preset in the camera. Since the observed materials are of different kinds and with different emissivity values for the current observation, the emissivity set in the camera is 0.86. Therefore, the temperature measured in the camera should only be seen only as a relative measurement. Thermal tests were carried out with the selected insulating layer, shown in Figure 12, in addition to only using an air gap for insulation, shown in Figure 13. An air gap can be considered as an interesting insulation layer as it reduces the complexity in the fabrication process as well as the weight. However, the thermogram does not show sharp temperature gradients which questions its usage as insulating "material" in this application.

IOP Conf. Series: Journal of Physics: Conf. Series **922** (2017) 012002 doi:10.1088/1742-6596/922/1/012002

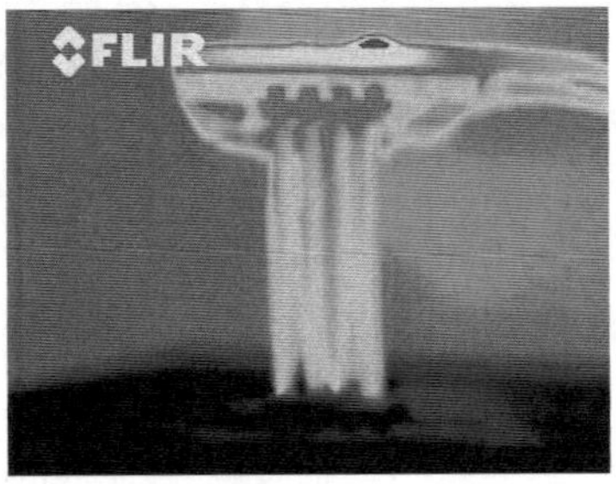

Figure 13. Multi stage thermal observation using IR camera (with air gap).

Figure 14. Bubbles leaked through the bulk material when the compressor was immersed in water and pressurized gas was passed into the inlet.

5.2. Buble Leak Test

The single stage compressor was leak tight and showed a leak rate of 10^{-2} mbarL/s which is nearly impossible to spot in the bubble leak test. The four-stage compressor enclosure was initially thought to be completely sealed after assembly. However, bubble leak testing by passing slightly pressurized gas across the compressor while it was immersed in water revealed this was not the case. Bubbles appeared around the compressor as shown in Figure 14. After several iterations of gluing over leaking areas it was concluded that the gas was leaking through the compressor enclosures bulk material (ABS plastic).

5.3. Performance Test

The single stage compressor produced very different pressure gradients for the various materials tested for the Knudsen compressor application. The performance of different membrane material is listed in Table 1. The most interesting material, Knudsen membrane, produced a pressure difference of 2100 Pa when subjected to a cold and hot temperature of 13°C and 99.9°C respectively. Due to the large amount of heating and cooling applied in the test setup most of the materials were observed to achieve a satisfactory temperature difference. The achievable pressure difference was however modest for all the material except two. The Knudsen membrane* showed the highest-pressure difference and the PMMA sheet the second highest pressure difference of only 150 Pa. PMMA was limited by the ability to surpass the gas. According to Bernoulli's principle when the flow rate increases, the pressure of the flow decreases. Due to this inverse proportionality of the flow rate and the pressure build up, the flow is stopped when PMMA sheet reaches the maximum pressure. Having the system at different pressure levels, it was inferred that the maximum compression was limited by leaks in the system.

The multi stage design allows for very simple variation in the number of connected steps. Three and four stage compressors were designed and fabricated. Initially the heater, the cooler, electronics and power supplies were turned on until the pressure of the compressor was stabilized. The data from the sensing elements were recorded and plotted in real time. Thermal transpiration was demonstrated over a period by varying the applied temperatures in the four-stage compressor. However, the temperature on the hot side of the compressor eventually melted the Knudsen membrane which in turn caused internal gas leaks that limited our success with the first multi stage compressor.

Table 1. Summary of performance characteristics for different membrane materials. Temperatures are in Celsius [7]

Material	$T_{cold}(°)$**C**	$T_{hot}(°)$**C**	$\Delta T(°)$**C**	$P_{max}(diff, Pa)$	flow (ml/min)
PI 0.02 μm	28	100.3	72.3	13	0.00
PI 0.2 μm	27.3	100.4	73.1	2.75	0.22
PI 1 μm	28.6	99.9	71.3	1.5	0.41
PMMA sheet	14.3	80.5	66.2	< 150	0.00
Silica aerogel	12	81.7	69.7	4	0.99
Open (ABS net)	12.2	82.6	70.4	0	0.00
Knudsen membrane*	**13**	**99.9**	**86.9**	**2100**	**1.2**
Thermal Wrap^{TM}	11.6	79.4	67.8	< 1	0.14
Spaceloft^{TM}	11.6	79.9	68.3	< 1	1.8

* Material used is confidential

The three-stage compressor was therefore maintained very carefully at 349 K and 283 K at the hot and the cold end respectively. The three-stage compressor achieved a pressure difference of 290 Pa. This is far less than the single stage compressor and we believe this was due to leaks in the compressor casings bulk material. However, when the temperature is varied, the pressure also changed in accordance to it, which indicates that the compressor is functional and behave more as expected.

6. Heat exchanger and JT valve combination

After the design and development of the Knudsen compressor, the next logical step is to integrate it with the other part of the JT cooling cycle, i.e. the heat exchanger and JT valve. Depending on the pressure ratio and the temperature required at the cold end, these parts may vary in dimension and specifications. However, with a holistic view, they were conceptually designed and tested.

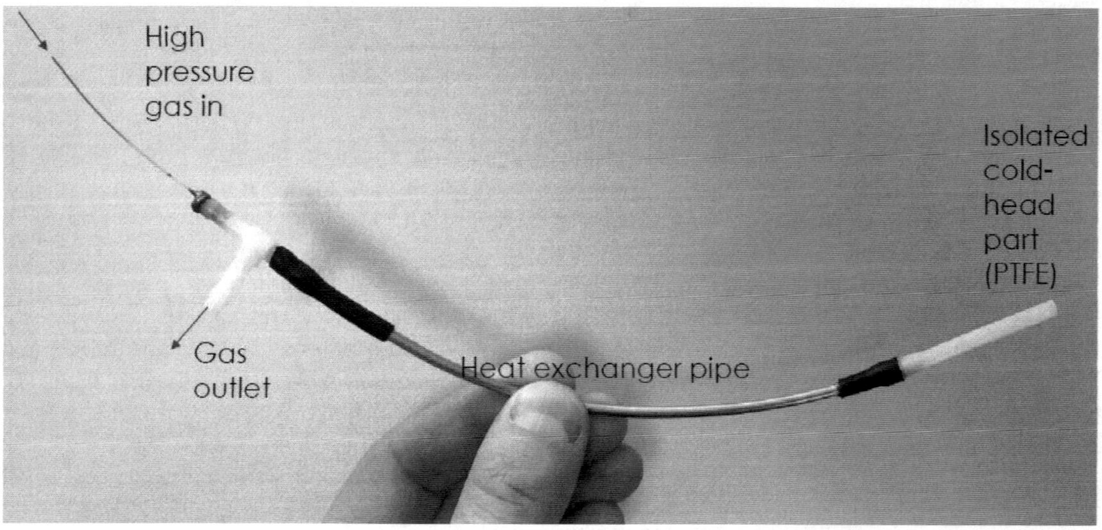

Figure 15. Heat exchanger and JT valve combination.

As a next step, a Tube-in-Tube heat exchanger with a length in the order of a few tenths of centimeters was designed and fabricated. The dimensions of the Tube-in-Tube heat exchanger were updated in system iterations to meet the JT cryocooler requirements and are kept confidential. While tested for short duration of less than an hour, with pressurized CO_2, the heat exchanger achieved a 7.5°C temperature decrease for a 5 bar CO_2 pressure. Figure 15 shows the miniature Tube-in-Tube heat exchanger that was developed and tested in the laboratory at APR Technologies. The test was conducted with the CO_2 gas and a prominent temperature was obtained at the cold end of the heat exchanger. A JT valve and thermally isolated volume are being developed to integrate a complete JT-cryocooler system

7. Results and Conclusions

The two key factors to increase the pressure ratio of the Knudsen compressor are the number of stages and the temperature difference across the Knudsen membrane. The number of stages of the compressor is limited by the weight as well as the complexity in fabrication. The temperature difference is also limited by the properties of the Knudsen membrane to take up the hot/cold temperatures without deformations. Three Knudsen compressors (Single stage, three stage and four stage compressors) were developed at APR Technologies and their performance were recorded. In all the three compressors, the GbF reached 90% of the applied hot/cold temperatures at the Knudsen membrane. Additionally, pressure gradients achieved by two of the compressors were recorded. The single stage compressor showed a good pressure difference of 2100 Pa while the three-stage compressor produced a pressure difference of only 290 Pa. The drastic pressure loss in the three-stage compressor was due to bulk gas leakage in the 3D printed ABS plastic case that was implemented for weight reduction and ease of manufacturing and assembly. The four-stage compressor developed a gross leak due to multiple thermal transpiration tests conducted on it, which prevented pressure measurements. The stage pressure ratio for each of the compressor were computed to 1:1.1. To achieve the JT cryocooler requirement of 1:10 pressure ratio, it is expected to have at least a 24-stage compressors. However, with design improvements we believe it is possible to reach the specified compression ration with far fewer stages. The Knudsen compressors developed at APR Technologies are light weight, do not contain any moving parts and consumes minimal power. For example, the four-stage compressor that was developed weighs less than 150 grams and consumes 12 Watts power to induce the thermal energy needed for its operation.

8. Future Work

First and the foremost, the compressor enclosure is to be replaced with non-leaking bulk material. This bulk material shall be molded plastic or a lightweight metal that can withstand the operating pressure range. Temperature limitation at the hot end to avoid melting of the Knudsen membrane is also foreseen. Further, the compressor will be integrated to the other company made parts of the cryocooler system. The fully developed cryocooler is expected to undergo normal space qualification tests including radiation, thermal cycling and vibration. Optimization is required to be done essentially in the functionality, cost, reliability, weight, lifetime and power requirements of the system.

Acknowledgments

Sincere thanks to the project sponsors VINNOVA and SIO GRAFEN for their faith on the APR team and investing on this valuable R&D project. Also, many thanks to Swedish National Space Board (Rymdstyrelsen) for funding the initial study that established a firm platform for this project.

References

[1] Sherman A 1983 *National aeronautics and space administration needs and trends in cryogenic cooling Cryogenics* **23** 348 - 352
[2] Donabedian M 2003 *Spacecraft Thermal Control Handbook: Vol 2 Cryogenics* (California: Aerospace Press)
[3] Rando N 2010 Cryogenics in space *Observing Photons in Space* p 593-606, (Springer)
[4] Derking J H, Holland H J, Tirolien T and Ter Brake H J M 2012 *A miniature joule thomson cooler for optical detectors in space* (Review of scientific instruments vol. 83 no. 4) p. 045117
[5] Maytal B and Pfotenhauer J M 2012 Miniature Joule-Thomson cryocooling: principles and practice (Springer Science & Business Media)
[6] Gupta Y B, Kumar N and Gianchandani 2010 A high-flow Knudsen pump using a polymer membrane: Performance at and below atmospheric pressures *IEEE Conf. 23 Micro Electro Mechanical Systems (MEMS)* 1095 - 1098
[7] Antelius M 2014 *Self-generating Pumped Meso/Micro-Cryo Cooling System without Moving Parts for Low Noise Applications* APR Technologies AB Internal company technical report,

IOP Conf. Series: Journal of Physics: Conf. Series **922** (2017) 012003 doi:10.1088/1742-6596/922/1/012003

A new approach for the calculation of falling droplets from a cylindrical glass capillary based on force balance and velocity

Sebastian Hummel, Martin Bogner, Michael Haub, Joachim Saegebarth and Hermann Sandmaier

University of Stuttgart - Chair Microsystems, Pfaffenwaldring 4F, 70569 Stuttgart, Germany

E-mail: sebastian.hummel@mst.uni-stuttgart.de

Abstract. This paper presents a new simple analytical method to estimate the properties of falling droplets without solving complex differential equations. The derivation starts from the balance of forces and uses Newton's second law and the equations of motion to calculate the volume of growing and detaching droplets and the time between two successive droplets falling out of a thin cylindrical capillary of borosilicate glass. In this specific case the reservoir is located above the capillary and the hydrostatic pressure of the fluid level leads to drop formation times about one second. In the second part of this paper experimental results are presented to validate the introduced calculation method. It is shown that the new approach describes the measuring results within a deviation of ±6.2%. The third part of the paper sums up the advantages of the new approach and an outlook is given on how the research on this topic will be continued.

1. Introduction

In today's microfluidic hanging and falling droplets are widely used, for example in medical dosing devices. These are used to measure small amounts of liquids for specific applications, for example in scientific laboratories for the preparation of different substances or in the clinical dosage of medication in infusions.

Tate [1] was the first to investigate the volume of falling droplets. He presented an approach starting from the balance of surface force and gravitational force to calculate the maximum mass of a falling droplet. Rayleigh [2] and Harkins & Brown [3] investigated the weight of falling droplets in their experiments. They discovered that the droplets are smaller than proposed by [1] and they expanded Tate's approach by experimental correction factors based on the specific capillary length. The aim of the approaches presented in [1], [2] and [3] was to estimate the maximum droplet size under quasi-static conditions.

More detailed approaches based on differential equations are presented by Chesters, Eggers & Dupont and Clanet & Lasheras. Chesters [4] derived several equations for different parts of the shape of a droplet which enables him to plot different states in the droplet formation process. Eggers & Dupont [5] used the Navier-Stokes-Equations in cylindrical coordinates to simulate the shape of growing, necking and detaching droplets numerically and Clanert & Lasheras [6] focused on the transition from dripping to jetting. The approaches [4], [5] and [6] are on the one hand much more accurate and on the other hand much more complicated than those presented in [1], [2] and [3].

Comprehensively none of the above presented approaches provides an easy and approximately accurate method to calculate falling droplets. This gap will be closed with this paper by presenting a new simple method based on the balance of forces acting on the hanging droplet to estimate the volume and the time between two successive falling droplets.

2. The new calculation approach

Starting from Tate's ideas [1] the balance of the forces acting on the droplet leads to an approach which can be divided up in three phases:

(i) In the first phase the upwards directed forces overcome the downwards directed forces and the droplet is hanging on the bottom of the capillary. The maximum size of a hanging droplet is reached in the equilibrium of forces.

(ii) In the second phase the downwards directed forces overcome the upwards directed forces. The resulting force accelerates the droplet in the direction of the gravitational constant and finally leads to the detachment in Phase 3. As long as the velocity of the droplet is slower than the velocity of the fluid streaming out of the capillary and into the droplet the growth of the droplet goes on, which is already described by Neumann & Seeliger [7].

(iii) If the velocity of the droplet overcomes the velocity of the fluid flow no more fluid can reach the drop and the falling drop has reached its maximum size. The fluid streaming out of the capillary until the final detachment of the droplet is already used to build a new droplet.

2.1. Phase 1

The force acting upwards on a hanging droplet is the surface force

$$F_{sf} = 2 \cdot \pi \cdot R \cdot \sigma \tag{1}$$

where R is the outer radius of the cylindrical capillary according to the observations presented in figure 2 and σ the surface tension. While the droplet grows hanging at the capillary, its shape changes from spherical to the typical shape of a hanging droplet and the surface force increases until a cylindrical transition from the capillary to the droplet is reached. This transition point represents a threshold which has to be overcome before the droplet detaches. Therefore the surface force is assumed to be maximum in equation (1).

The first force acting downwards on a hanging droplet is the gravity

$$F_g = \rho \cdot V \cdot g \tag{2}$$

where ρ is the density, V the volume of the hanging droplet and g the gravitational constant. The second force is the pressure force

$$F_p = p \cdot A_{cap} \tag{3}$$

due to the pressure p in the droplet acting on the plne surface of the capillary

$$A_{cap} = \pi \cdot \left(R^2 - r^2\right) \tag{4}$$

which is covered by the droplet, with R being the outer radius of the capillary and r being the inner one. The capillary is assumed to be hydrophilic and the edge between the plane surface and the curved surface of the capillary is assumed to act as an geometrical stop for the fluid. Due to its immobility the capillary responses to this force by an opposite directed force which is considered here as a force acting downwards on a hanging droplet. Compared to [1] the pressure force is added in this approach and explains the smaller droplets detected by [2] and [3]. For further simplification the pressure p inside the droplet is assumed to be constant over the droplet formation process though it changes with the increasing droplet volume according to the law of Young-Laplace [8]. To calculate the droplet properties an average pressure is used, computed with the maximum droplet pressure at the given capillary and the pressure in the droplet at the equilibrium of forces, assuming the shape of the droplet to be spherical. According to this assumptions the droplet volume is the only parameter changing with time and the equilibrium of forces can be used to calculate the droplet volume V_1 in the equilibrium state by an iterative procedure.

$$V_1 = \frac{F_{sf} - F_p}{\rho \cdot g} = \frac{2 \cdot \pi \cdot R \cdot \sigma - p \cdot A_{cap}}{\rho \cdot g} \tag{5}$$

Using the volumetric flow rate $\dot{V_0}$ into the droplet, which can be calculated by the law of Hagen-Poiseuille [9], the time t_1 needed to reach the equilibrium state starting from a completely filled capillary can be calculated.

$$t_1 = \frac{V_1}{\dot{V_0}} \tag{6}$$

2.2. Phase 2
Phase 2 starts from the equilibrium state in which the droplet has already reached the volume V_1. Due to the ongoing flow into the droplets its volume continues increasing and the downwards directed forces starts to overcome the surface force. According to Newton's second law the resulting acceleration

$$a = \frac{dv}{dt} = \frac{F_g + F_p - F_{sf}}{m} \tag{7}$$

of the droplet in the direction of the gravitational constant can be calculated with the mass m of the hanging droplet. The integration of equation (7) leads to the droplet velocity v depending on the time t.

$$v = g \cdot t - \frac{F_{sf} - F_p}{\rho \cdot \dot{V_0}} \cdot ln\left(1 + \frac{t}{t_1}\right) \tag{8}$$

Computing the intersection of v with the inlet velocity v_0 leads to the time t_2 which describes the time elapsing in phase 2. t_2 enables the calculation of the volume V_2 for which the droplet volume is increasing in phase 2.

$$V_2 = \dot{V_0} \cdot t_2 \tag{9}$$

2.3. Phase 3
Phase 3 deals with the final detachment of the droplet. As already described no further increase of the droplet volume has to be expected. An appraisal for the necking time t_3 is given by Eggers [10] and modified with experimental results by Clanet & Lasheras [6]. Using Eggers' original approach

$$t_3 = \sqrt{\frac{r^3 \cdot \rho}{\sigma}} \tag{10}$$

with r being the radius of the cylindrical fluid jet or the neck of a droplet before the constriction begins, ρ being the density and σ being the surface tension, leads to a necking time about several milliseconds.

2.4. Summary
Summarizing the presented approach leads to the droplet volume

$$V_{droplet} = V_1 + V_2 \tag{11}$$

which consists of the volume V_1 for which the droplet is increasing in phase 1 and of the volume V_2 for which the droplet is increasing in the second phase.

The time

$$T = t_1 + t_2 + t_3 \tag{12}$$

is needed to form the first droplet until its final detachment out of an completely filled capillary. To calculate the time between two droplets it has to be mentioned that the fluid streaming through the capillary in phase 3 already contributes to the volume of the next droplet. Because the calculation of t_1 starts from a completely filled capillary t_3 has to be subtracted from t_1 to calculate the time ΔT between two successive droplets.

$$\Delta T = (t_1 - t_3) + t_2 + t_3 = t_1 + t_2 \tag{13}$$

3. Validation of the calculation method

3.1. Experimental setup

To validate the new calculation approach a measurement setup was created using a RaspberryPi microcontroller and a RaspberryPi camera module. This setup allows to record the droplet formation and detachment with up to 118 frames per second and a resolution of 640 x 480 pixels, slightly reduced because of the internal processing of the images by the RaspberryPi microcontroller. The sample with the capillary to test and the reservoir located above the capillary are fixed by an tripod clamp and illuminated by a LED with the power of 1 W. The RaspberryPi microcontroller is programmed using python and OpenCV and allows an automated evaluation of the recorded droplets using a graphical user interface. The setup is shown in figure 1, an exemplary droplet of water near its point of detachment is shown in figure 2.

Figure 1. The figure shows the measurement setup with the camera module in the blue box on the left, the cylindrical capillary and the spot of the LED on a diffuser on the right.

Figure 2. Photography of a droplet hanging on a cylindrical capillary near its point of detachment (Phase 3).

The experimental setup consists of a cylindrical capillary of borosilicate glass with an outer diameter of 880 μm, an inner diameter of 440 μm and a length of 20 mm. Different fluid levels, measured over the bottom of the capillary are used to vary the hydrostatic pressure acting on the flow through the capillary and setting the volumetric flow rate $\dot{V}_0$ due to the law of Hagen-Poisseuille [9]. In scope of this work four different levels between $h = 37.5$ mm and $h = 63$ mm with an accuracy of ± 0.25 mm have been used, each with 17 to 70 analyzed droplets to satisfy statistical methods. The fluid used in the experiments is degassed water. Due to the significant change of the properties of water in the range of room temperature [11], the temperature of the fluid was monitored and the obtained information were used to correct the properties of water used in the calculations by linear interpolation according to the temperature at the time of the experiments.

Although the frame rate and the resolution of the camera are low compared to an industrial high-speed-camera, the described measurement setup provides accurate results. Inaccuracies can occur for example due to the required vertical orientation of the sample which has to be done manually by the operator. This has to be mentioned while interpreting the measurement values.

3.2. Results and discussion

Figure 3 and 4 show the results of the measurements and the calculations (dashed line). The error indicators at the measured values show the standard deviation of the measurements. The accuracy of the presented calculation is represented by the maximum deviation between the calculated and the measured values of ±6.2%.

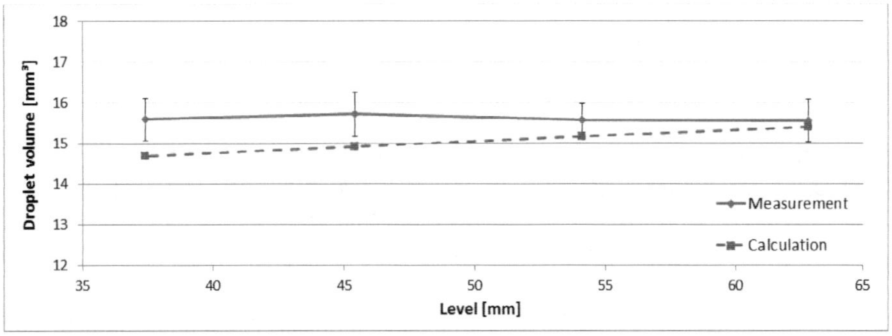

Figure 3. The figure sums up the measured droplet volumes for different fluid levels in the reservoir and compares them to the calculated values (dashed line).

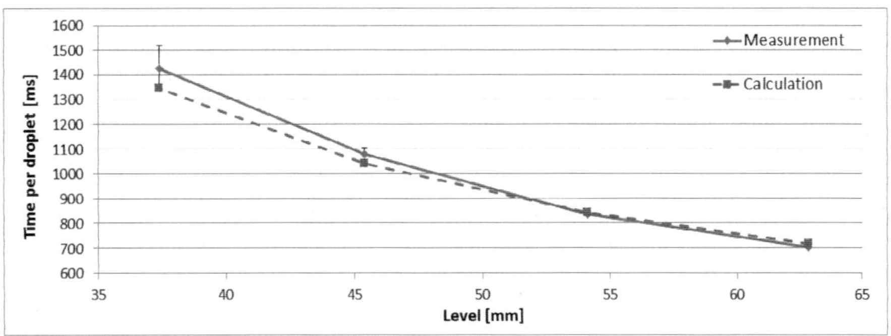

Figure 4. The figure compares the measured and calculated time between two successive droplets for different fluid levels in the reservoir. It shows the decreasing time needed to form a droplet with increasing hydrostatic pressure acting on the capillary.

Figure 3 and 4 validate the presented calculation approach. The accuracy of the measurement setup is shown by the standard deviations which are with a maximum value of 7% of the average values small compared to the average values.

The measurement values fit to a ±6.2% belt around the calculation results. Merely the values for the levels $h = 37.5$ mm and $h = 45.5$ mm in figure 3 differ from the calculated values for these fluid levels. This could be explained by the difficulties which occurred while recording the measurement values: The software analyzing the recorded pictures like the one shown in figure 2 has to walk through several steps to divide the capillary (the part of the picture which is not changing from one picture to the next) of the droplet. Due to the slow growth of the droplet the recognition of the capillary is partially incorrect and therefore the measurement results are also incorrect. The second thing to be mentioned regarding figure 3 is the trend shown by the curves. While the calculated values increase with the fluid level, the measured values stagnate at a fixed level. A final explanation for this phenomena can not yet be given, it might be explained by inaccuracies of the measurement or the calculation but further investigations are required.

4. Conclusions

In this paper a new approach was presented to estimate the droplet volume and the time between two successive droplets. The approach is validated by experimental results and corresponds to the accuracy of ±6.2%. This shows that it is in principle possible to estimate the droplet volume and the time between two successive droplets of dynamic droplet formation processes with little effort and acceptable accuracy without using complex differential equations.

5. Outlook: additional investigations done and further research topics

Though many insights have been gained so far, there are lots of topics which has to be investigated. There are some phenomena in figure 3 which could not be explained so far and which have to be explained in further investigations. Changing the geometric parameters of the capillary will change the droplet formation process as well as changing the capillary material or the fluid used to form the droplets, for example due to hydrophilic or hydrophobic behavior of the material-fluid-combination. Using glycerol in the same capillaries as described in this work, for example, leads to droplet formation times in the scale of minutes.

In addition to the described additional investigations an extension of the range of fluid levels which leads to different flow rates into the droplet has to be investigated to get the validity area of the presented approach. To complete the validation of the presented approach numerical simulations have to be performed, for example using the CFD-tool OpenFOAM.

Finally it has to be mentioned that Neumann & Seeliger [7] detected a maximum droplet size in their experiments where the droplet volume stops increasing with increasing flow rate into the droplet and starts decreasing again due to a further increase of the flow rate. This effect is not yet considered in the presented approach and has to be taken into account in future investigations, maybe being responsible for the phenomena in figure 3.

References

[1] Tate T 1864 *The London, Edinburgh, and Dublin Philosophical Magazine and Journal of Science* **27** 176–180
[2] Rayleigh L 1899 *The London, Edinburgh, and Dublin Philosophical Magazine and Journal of Science* **48** 321–337
[3] Harkins W D and Brown F 1919 *Journal of the American Chemical Society* **41** 499–524
[4] Chesters A K 1977 *Journal of Fluid Mechanics* **81** 609–624
[5] Eggers J and Dupont T F 1994 *Journal of Fluid Mechanics* **262** 205–221
[6] Clanet C and Lasheras J C 1999 *Journal of Fluid Mechanics* **383** 307–326
[7] Neumann H and Seeliger R 1939 *Zeitschrift für Physik* **114** 571–578
[8] Butt H J, Graf K and Kappl M 2003 *Physics and chemistry of interfaces* (John Wiley & Sons)
[9] Sutera S P and Skalak R 1993 *Annual Review of Fluid Mechanics* **25** 1–20
[10] Eggers J 1997 *Physik Journal* **53** 431–434
[11] Haynes W M (ed) 2016 *CRC Handbook of Chemistry and Physics* vol 97 (Internet Version 2017) (CRC Press/Taylor & Francis, Boca Raton, FL)

Design, fabrication and characterisation of a microfluidic time-temperature indicator

P Schmitt[1], **K Wedrich**[1], **L Müller**[1], **H Mehner**[1,2] **and M Hoffmann**[1]

[1] Technische Universität Ilmenau, Ilmenau, Germany

[2] now with X-Fab MEMS Foundry GmbH, Erfurt, Germany

philip.schmitt@tu-ilmenau.de

Abstract. This paper describes a concept for a passive microfluidic time-temperature indicator (TTI) intended for intelligent food packaging. A microfluidic system is presented that makes use of the temperature-dependent flow of suitable food ingredients in a microcapillary. Based on the creeping distance inside the capillary, the time-temperature integral can be determined. A demonstrator of the microsystem has been designed, fabricated and characterised using liquid sugar alcohols as indicator fluids. To enable a first wireless read-out of the passive TTI, the sensor was read out using a commercial RFID equipment, and capacitive measurements have been carried out.

1. Introduction

One of the current global challenges is the waste of perishable goods. Recent European studies show that 20 % of the total EU food supply is being disposed although it is still esculent. This averages out at an annual food waste of about 173 kg per person [1]. The most important reason for discarding food in the private and reselling sectors are exceeding expiration dates that are usually printed on food packages. However, since the shelf-time of perishable goods depends on environmental conditions such as temperature and humidity, the labelled expiration date on packages can be seen as a cautiously estimated reference which does not strictly reflect the actual condition of the enclosed item.

In order to extend shelf-time of food, research on intelligent packaging has strongly increased, recently [e.g. 2]. Intelligent packaging systems focus on the replacement of the printed expiration date by the integration of passive sensors that measure and indicate the current spoilage condition of the enclosed perishable item.

Since quality deterioration of food is strongly impaired by elevated temperatures and the time the food is exposed to them, the integral of a time-dependent temperature profile can be used as an indicator for the condition of a perishable item. Passive time-temperature indicators (TTI) are therefore often used as monitoring devices on food packages. Existing and commercially available TTIs are typically based on chemical reactions, polymerisation or diffusion processes [3, 4]. By changing their colour, these non-electrical tags optically indicate the current state of the product.

However, a pure optical evaluation of the measurement makes it difficult to calculate and predict a new and adapted expiration date. Moreover, such sensors are usually placed at the outside of the package. Furthermore, an automated and electronic read-out of the sensor during handling or transportation within the supply-chain may be difficult by visual inspection, only.

IOP Conf. Series: Journal of Physics: Conf. Series **922** (2017) 012004 doi:10.1088/1742-6596/922/1/012004

In this paper we propose a new, optimised passive microfluidic TTI-system based on the works presented in [5]. The TTI-system can be connected to an RFID transponder enabling a wireless read-out. The proposed system is based on the progressive filling of a capillary by a dielectric fluid. Since the viscosity of the fluid is temperature-dependent, the filling speed of the capillary strongly depends on the imposed temperature. The covered distance of the fluid in the capillary consequently corresponds to the time-temperature integral. By employing an indicator-fluid featuring a high dielectric constant, the creeping distance can be determined via a capacitive measurement that enables an electrical evaluation of the sensor. In the framework of this paper, we focus on the evaluation of appropriate indicator fluids and their fluidic properties in a microcapillary. Fluids are stringently chosen from food ingredients: non-toxic and responsible for the aging of perishable food.

2. Method

The working principle of the microfluidic TTI is shown in figure 1. The main element of the sensor is a capillary with rectangular cross-section. Along the capillary, two electrodes are placed on opposite side walls so that a plate-capacitor is formed. Additionally, a reservoir filled with an indicator fluid is positioned at the beginning of the capillary. Depending on the considered indicator material, it is initially either in a highly viscous or in a solid state. With rising temperature, the indicator material starts to melt, becomes less viscous and starts to creep into the microcapillary. As the temperature lowers, the viscosity increases again, forcing the fluid to decelerate and stop. As the creeping distance depends on both the imposed temperature and the exposure time, the final creeping distance of the fluid can be seen as a measure of the time-temperature integral.

To read out the final creeping distance electronically, the capacity between the side-wall electrodes is measured. Since the indicator fluid features a higher dielectric constant than air, the capacity of the plate capacitor increases almost linearly with the creeping distance.

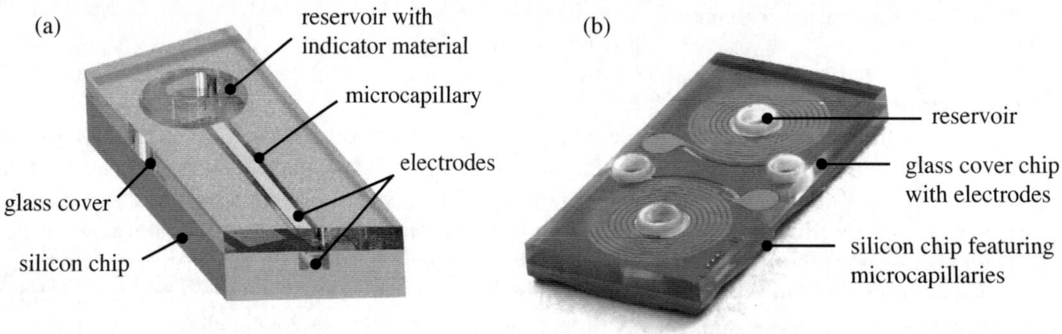

Figure 1. (a) Schematic illustration of the TTI; (b) fabricated microfluidic TTI demonstrator

2.1. Microfluidic model

To predict the dynamic behaviour of the fluid and the creeping distance x within the capillary, a microfluidic model based on the Navier-Stokes-Equation is derived. All significant forces that apply are visualized in figure 2. Taking the forces into account, a one-dimensional formulation of the Navier-Stokes equation can be given as:

$$\frac{\mathrm{d}}{\mathrm{d}t}\left(m\frac{\mathrm{d}x}{\mathrm{d}t}\right) = -F_{\mathrm{weight}} - F_{\mathrm{vis}} + F_{\mathrm{capillary}} \tag{1}$$

Adapting the individual force-terms for a rectangular cross-section, the above equation can be reformulated to:

$$\varrho bhx\frac{\mathrm{d}^2x}{\mathrm{d}t^2} + \varrho bh\left(\frac{dx}{dt}\right)^2 = -\varrho bhgx - \zeta(n)\eta\, x\frac{\mathrm{d}x}{\mathrm{d}t} + 2\sigma_{\mathrm{lg}}\cos\theta(b+h) \tag{2}$$

where ϱ is the density and η the dynamic viscosity of the fluid. σ_{lg} represents the fluid surface tension and θ is the contact angle as shown in figure 2. $\zeta(n)$ substitutes the following geometric function respecting the width b and the height h of the capillary:

$$\zeta(n) = \frac{12\,(bh)^2}{\min(b,h)^3\max(b,h)\left(1-\sum_{n=1}^{\infty}\frac{1}{(2n-1)^5}\frac{192}{\pi^5}\frac{\min(b,h)}{\max(b,h)}\tanh\left((2n-1)\frac{\pi\max(b,h)}{2\min(b,h)}\right)\right)} \quad (3)$$

Since the diameter of the microcapillaries does not exceed the order of several microns, scaling effects have to be taken into account. After analysing the characteristic Bond and Weber numbers, the inertial force and the fluid's weight can be neglected in comparison to the capillary force, allowing a simplification of equation (2):

$$\zeta(n)\eta\, x\frac{\mathrm{d}x}{\mathrm{d}t} = 2\sigma_{\mathrm{lg}}\cos(\theta)(b+h) \quad (4)$$

An analytical solution of the above differential equation can be found as:

$$x(t,T) = 2\left(t\,\frac{(b+h)\,\sigma_{\mathrm{lg}}(T)\cos(\theta(T))}{\zeta(n)\;\eta(T)}\right)^{\frac{1}{2}} \quad (5)$$

It is important to note that almost all material parameters are temperature-dependent, thus enabling the evaluation of the creeping distance x not only as a function of time, but also as a function of temperature.

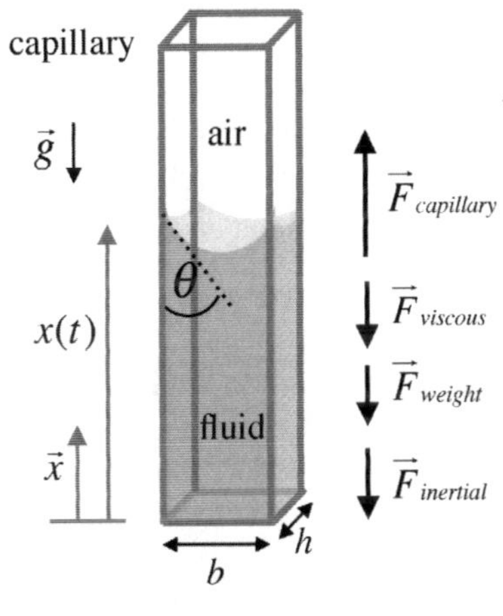

Figure 2. Illustration of relevant forces and geometric parameters in a vertically oriented capillary.

2.2. *Capacitive measurement of the creeping distance*

For the electrical read-out of the time-temperature integral, the associated creeping distance of the fluid has to be translated into an electrical signal. By using an indicator fluid with a high dielectric constant $\varepsilon_{r,fluid}$, the evaluation of the creeping distance can be achieved by measuring the capacity change of the plate capacitor. The total capacity as a function of the creeping distance can be approximated by:

$$C_{\mathrm{p}}(x) = \varepsilon_0\frac{b}{h}(L+x(\varepsilon_{\mathrm{r,fluid}}-1)) \quad (6)$$

where L is the length of the capillary and ε_0 is the vacuum permittivity.

3. Indicator material

The choice of a suitable indicator fluid is essential for a proper operation mode of the TTI. Many restrictions regarding fluidic and dielectric properties as well as safety and health issues have to be considered.

Since different perishable items are sensitive to different storage temperatures, the melting temperature or rather the glass transition temperature of the material should be adjustable, facilitating an easy adaption of the TTI for a wide range of products. Moreover, fluidic requirements such as a low contact angle or a high fluidic surface tensions must be fulfilled to enable a strong capillary force. In order to allow an increased sensitivity to capacitive changes, the indicator material should also feature a high dielectric constant and a low electrical conductivity. Furthermore, for being integrated in food packages, the fluid must not be toxic. Food ingredients are surely non-toxic and first choice.

Therefore, sugar alcohols like sorbitol or xylitol are ideal fluids to serve as indicator materials in the TTI-sensor. These food additives are usually used as artificial sweeteners in many food products and can be considered as harmless chemical products. Because of their polar molecular structure, sugar alcohols normally offer high dielectric constants what makes them suitable for capacitive measurements.

Another advantage is that they can appear either in a crystalline or in an amorphous state. Whereas the melting temperature of crystalline sorbitol was found to be 95 °C, the glass temperature of amorphous sorbitol is known to be –2.3 °C [6]. Therefore, and using the respective form of sugar alcohol, a wide range of temperatures can be covered. Furthermore, by dissolving sugar alcohols in water, the glass transition of the dissolution can be varied what allows an adaption of the indicator material for different temperature ranges as well. Moreover, the mobility of sugars and fats are. responsible for the aging of ice cream and other products in an insufficiently cold storage.

Since sugar alcohols have not been considered as indicator material in a microfluidic sensor before, various dielectric and microfluidic experiments have been performed in order to characterise the material properties of molten sorbitol and xylitol. Pure D-sorbitol and xylitol purchased from Sigma-Aldrich were used for experimental investigations.

3.1. Permittivity measurement

Firstly, the frequency-dependent dielectric constant of amorphous sorbitol has been measured. Therefore, a cylindrical test-capacitor made of anodised aluminium has been used. The capacity of the *empty* capacitor has been measured utilizing an *Agilent E4980A* LCR-meter first. Subsequently, the test-capacitor was filled with molten amorphous sorbitol, and the capacitance measurement was repeated at a temperature of 40 °C. Another measurement at room temperature was repeated 24 hours later after the sorbitol had recrystallized.

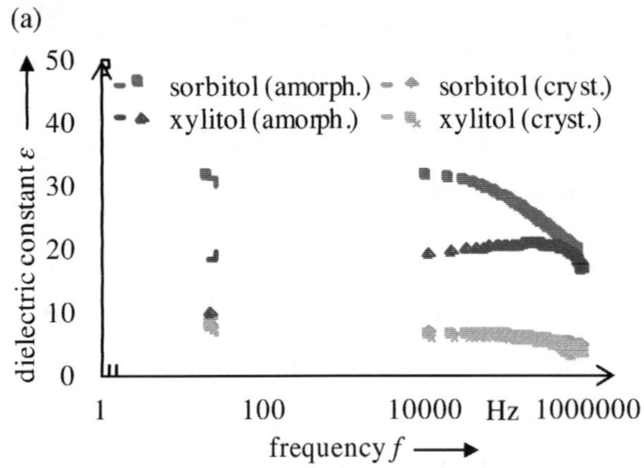

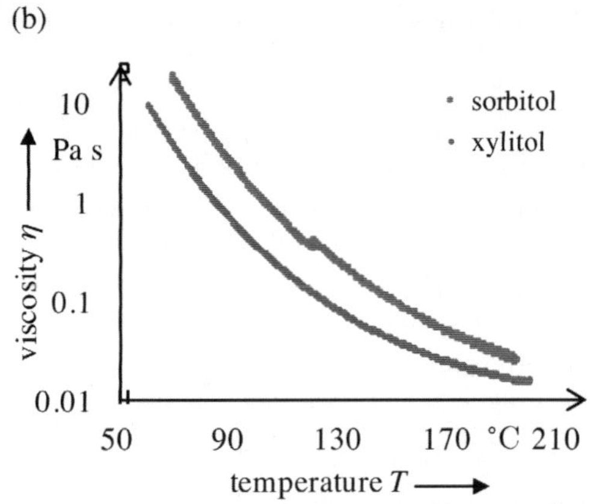

Figure 3. (a) Measured dielectric constant of sorbitol and xylitol in crystalline and amorphous state; (b) Temperature-dependent viscosity measurement of sugar alcohols.

The quotient of the measured capacitance of the filled and empty capacitor finally yields the dielectric constant of the sugar alcohol. The results of the measurements are shown in figure 3a. The difference of the dielectric constant between amorphous and crystalline sugar alcohol can be led back to the different molecular mobility in crystalline and amorphous state.

3.2. Viscosity measurement

The most significant property which determines the temperature-dependent creeping speed of the fluid inside a capillary is the viscosity. Dynamic viscosity measurements were performed utilizing an *Anton Paar MCR 302*. Since the sugar alcohols were initially in a crystalline state, they have been molten first at high temperatures. The viscosity has then been measured for declining temperatures. The measurement results for sorbitol and xylitol are presented in figure 3b.

4. Device fabrication

A first microfluidic demonstrator of the TTI-system has been fabricated using standard MEMS technologies. Microcapillaries of different sizes with rectangular cross-sections are etched into a silicon wafer by deep reactive ion etching (DRIE). In order to save space but to ensure a homogenous fluidic flow at the same time, a helical arrangement of the capillaries was chosen as shown in figure 1b. The capillaries are covered by a borosilicate glass wafer which is bonded to the silicon wafer. Both the silicon substrate and the glass cover feature metallic electrodes along the capillary that form a plate capacitor. To enable an external electrical connection to the electrodes and to provide an inlet for the indicator fluid, holes are laser drilled into the glass cover.

5. Fluidic characterisation

For a microfluidic characterisation of the temperature- and time-dependent creeping distance, the experimental setup as shown in figure 4a has been used. Demonstrator chips were annealed to stationary specific temperatures, using a regulated hotplate. When the respective chip has reached the final temperature, crystalline sugar alcohol was filled into the reservoir of the demonstrator. The immediately starting creeping process has been measured with a known framerate so that the filling speed could be evaluated. Figure 4b shows the measurement results for the time-dependent creeping distance of sorbitol for two different temperatures in a 100 μm x 100 μm silicon capillary. The associated theoretical graphs in figure 4b are based on equation (5).

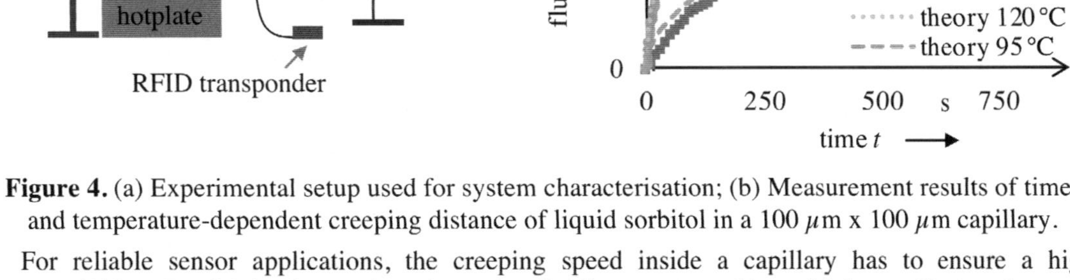

Figure 4. (a) Experimental setup used for system characterisation; (b) Measurement results of time- and temperature-dependent creeping distance of liquid sorbitol in a 100 μm x 100 μm capillary.

For reliable sensor applications, the creeping speed inside a capillary has to ensure a high reproducibility. By fabricating and testing microfluidic systems with multiple capillaries that start from the same reservoir, a high reproducibility could be verified as indicated in figure 5a. The presented x-t-profiles refer to the average creeping distance in six capillaries of the respective chip.

In order to test a first electronical wireless read-out of the passive TTI-sensor, the microfluidic system was connected to a capacity evaluation circuit based on commercial RFID equipment. As output, the RFID evaluation circuit provides a voltage which is proportional to the creeping distance. Measurement results, given in figure 5b, provide the proof of principle for capacitive read-out.

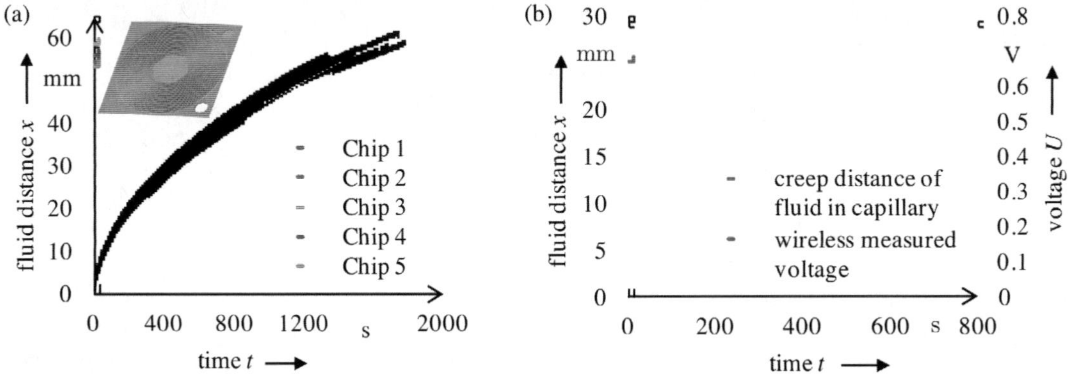

Figure 5. (a) Reproducibility testing using microsystems with multiple capillaries. (b) Electrical wireless read-out by RFID providing a corresponding voltage as reference for the creeping distance.

6. Conclusion

Non-electrical TTI-systems have been considered for a long time as reliable indicator for passive temperature surveillance of cool-chains or sterilisation processes. However, existing TTIs often allow only an optical evaluation of the measurement results. Furthermore, their adaption to different temperature ranges is limited. In this paper, a new microfluidic TTI concept has been presented that complements these missing features of today's commercially available TTI-solutions. The microfluidic system can be read out wirelessly using RFID technology.

The utilisation of pure sugar alcohols as indicator material in the TTI system has been assessed. Special attention was paid to their fluidic and dielectric properties. Measurement results show that sorbitol and xylitol offer great potential as indicator materials for high temperature applications like sterilisation processes. For low temperature applications such as cool-chain surveillance, water-based dissolutions of sugar alcohols could be applied. Further investigations and a fluidic characterisation concerning sugar alcohol dissolutions are pending.

References

[1] Stenmarck A s, Jensen C, Quested T, Moates G, Buksti M, Cseh B z, Juul S, Parry A, Politano A and Redlingshofer B 2016 *Estimates of European food waste levels*: IVL Swedish Environmental Research Institute)

[2] Fuertes G, Soto I, Carrasco R, Vargas M, Sabattin J and Lagos C 2016 Intelligent Packaging Systems: Sensors and Nanosensors to Monitor Food Quality and Safety *Journal of Sensors* **2016**

[3] Biji K, Ravishankar C, Mohan C and Gopal T S 2015 Smart packaging systems for food applications: a review *Journal of food science and technology* **52** 6125-35

[4] Kerry J, O'grady M and Hogan S 2006 Past, current and potential utilisation of active and intelligent packaging systems for meat and muscle-based products: A review *Meat science* **74** 113-30

[5] Schneider M and Hoffmann M 2011 Non-electrical-power temperature-time integrating sensor for RFID based on microfluidics. In: *SPIE Microtechnologies*: International Society for Optics and Photonics) pp 80661V-V-6

[6] Gombás Á, Szabó-Révész P, Regdon G and Erős I 2003 Study of thermal behaviour of sugar alcohols *Journal of thermal analysis and calorimetry* **73** 615-21

IOP Conf. Series: Journal of Physics: Conf. Series **922** (2017) 012005 doi:10.1088/1742-6596/922/1/012005

Time-resolved measurements of DNA interactions in an electrowetting-on-dielectric system using confocal microscopy

Lorenz T. Sparrenberg, Benjamin Greiner, Christian Mueller and Harald P. Mathis

Fraunhofer Institut for Applied Information Technology FIT; BioMOS; Schloss Birlinghoven, D-53754 Sankt Augustin, Germany

E-mail: lorenz.sparrenberg@fit.fraunhofer.de

Abstract. To identify new drug candidates a deep and profound knowledge of molecule interactions is needed. In the current work, a combination of an electrowetting-on-dielectric (EWOD) system with a confocal microscopy is presented for the first time. The aim of this research is to attain time-resolved information about nucleic acid interactions at a single molecule level. Confocal microscopy is a promising technique for reaction analysis on a molecular scale. But the liquid handling of the needed tiny volumes of highly diluted solutions is very challenging. An EWOD based system for droplet handling can address this demand. In this paper the development of an EWOD system for droplet handling in nanoliter scale is discussed and the combination of the EWOD system with a confocal microscope, to investigate nucleic acid interactions, is evaluated.

1. Introduction

Time-resolved information about molecular interactions are very important in biochemistry and pharmacology. To identify new drug candidates a deep and profound knowledge of drug-target interactions is needed. Methods based on fluorescence-fluctuation-spectroscopy (FFS) has become a standard in the analysis of molecular processes [1]. FFS is based on fluorescence fluctuations of molecules in tiny detection volumes [2]. A combination of confocal microscopy and improved photo detectors allows the detection of photon events originating from single molecules in highly diluted solutions [3]. To investigate time-resolved interactions in highly diluted nucleic acid solutions the liquid handling becomes a severe challenge. Tiny amounts of reagent solution have to be placed in the focus of a microscope at a defined starting point. By means of manual pipetting it is almost impossible to execute this task. Therefore a precise and reproducible handling technique for small liquid amounts in nanoliter scale is needed. An electrowetting-on-dielectric (EWOD) system for liquid-handling can address these demands. Electrowetting means to exert forces on phase interfaces such as liquid-gas or liquid-liquid interfaces by electric fields and was first discribed by Lippmann in 1875 [4]. This effect can be used to transport small droplets over an electrode array [5]. In recent years EWOD systems have drawn a lot of attention because they enable the manipulation of droplets in nanoliter scale [6]. Fundamental microfluidic operations with droplets have been demonstrated with EWOD systems (i.e. droplet transport, droplet splitting, droplet dispensing from reservoir) [7]. It has been shown that EWOD systems can transport not only pure water but also protein solutions [8], nucleic acid solutions [9] and even entire mammalian cells [10]. In the current work an EWOD system for droplet manipulation is developed and its potential to investigate time-resolved molecule interactions in combination with a confocal microscope is evaluated.

2. Experimental details

2.1. EWOD device fabrication

For the production of the electrowetting-on-dielectric (EWOD) device standard microfabrication techniques are used. The processing steps are displayed in figure 1. Both quartz and silicon wafers are used as substrates. Silicon wafers have excellent properties for photolithographic processes but because of their strong reflective behaviour fluorescence measurements with high signal to noise ratio are difficult. Hence, quartz wafers are employed for the combination of EWOD and confocal microscopy. Electrodes are formed by means of photolithography and consist of 20 nm chromium and 70 nm platinum. Reservoirs are situated ontop of the large

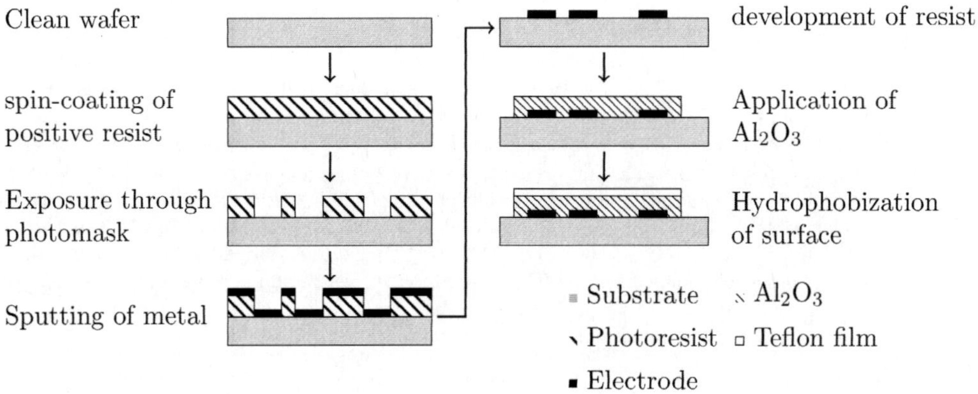

Figure 1: Processing of the EWOD structure. Standard photolithography methods were used. Al_2O_3 is applied by employing a sol-gel procedure. Tape is adhered onto the contact electrodes to prevent the formation of a dielectric layer ontop of them

electrodes on the left and the right side. From there small droplets can be taken. The electrode array at the bottom of the mask is used to contact the electrodes to the voltage supply. To reduce the number of contact electrodes some electrodes are linked together. As dielectric layer aluminum oxide is deposited using a sol-gel technique [11]. After a final tempering of the microfluidic structure at 450 °C a homogenous crack-free aluminum oxide layer of 400 nm thickness is obtained. The deposited aluminum oxide layer is smooth with no continuous pores. For guaranteeing a defined distance between electrodes and cover lid a frame of SU-8 with a thickness of 100 µm is deposited. For filling the reservoirs with liquid fine holes are bored with an ultrasonic drill. A glass plate with an indium tin oxide (ITO) layer is used as a cover lid and simultaneously as a ground electrode. Finally, an approximately 50 nm thick hydrophobic layer of teflon is spin coated on both the structure and the cover lid. To improve quality assurance interferometry measurements are taken from the final structure. The assembled device with a 3D-printed housing can be seen in figure 2. The needed voltage square wave is generated by a function generator and subsequently amplified to the required voltage level. An Arduino controlled in-house build circuit switches the voltage signal between adjacent electrodes. Non-activated electrodes are connected to ground. The circuitry is designed to cover a voltage range up to ±175 V. EWOD experiments were perfomed with a voltage of 80 V_{pp} with a frequency of 250 Hz (square wave). The droplets consist of DI-Water or 10 ng/µl DNA solution with 0.01 wt%

IOP Conf. Series: Journal of Physics: Conf. Series **922** (2017) 012005 doi:10.1088/1742-6596/922/1/012005

Figure 2: Photography of the EWOD bracket. The microfluidic tubes for probe injection and filling are on the left and right side. The connections for the circuitry can be seen at the bottom side.

TWEEN® 20 to reduce unspecific biomolecule adsorption to the teflon surface [9]. N-dodecane is used as ambient fluid to reduce friction forces and to stabilize the cassie-baxter state of the droplet on the surface and thus reducing contact angle hysteresis [12, 13].

2.2. Combination of EWOD device and microscope

The evaluation of the assembled EWOD device and microscopy system is conducted in two steps. A first simplified experiment is performed as a proof of concept. In this first experiment the number of involved components is reduced to a necessary minimum. Two droplets are merged by means of EWOD. One droplet consists of DI-water and the other one contains the fluorescence dye Alexa488 (1×10^{-9} M). Additionally, no ambient fluid but air is used in the EWOD system to avoid liquid-liquid interfaces that might influence the measurement. The water droplet is firstly moved to the detection spot. After this the droplet with Alexa488 dye is merged with the previous droplet. For fluorescence measurement a 63x oil emulsion objective (Zeiss) is used. Fluorescence excitation is done at 488 nm wave length.
In the next evaluation step a small droplet of DNA solution (10 ng/µl DNA with 0.01 wt% TWEEN® 20, 1:1000 Sybr Green I) is moved in and out of the detection spot of the microscope. The purpose of this evaluation is to point out the reproducible movement of nucleic acid solutions by means of electrowetting while measuring the fluorescence signal.

3. Results and discussion

3.1. Evaluation of EWOD device for liquid handling

To monitor the quality of the EWOD device interferometry measurements of the electrode structures are taken. The interferometry measurements ensure the integrity of the structures and the overlaying dielectric layer, an exemplary measurement is displayed in figure 3.

As it can be seen there are no cracks or obvious pores in the dielectric layer. The height difference between electrode and substrate originates from the different reflective properties of platinum and quartz that influence the measurement of the transparent Al_2O_3 layer. The functionality of the EWOD device is tested by a series of droplet manipulation experiments. A sequence of droplet displacements can be seen in figure 4. A solution of 10 ng/µl DNA solution with 0.01 wt% TWEEN® 20 is used. The switching time of the voltage is 500 ms. **A)** A 300 nl droplet is taken from a reservoir at the bottom of the image. **B)** The droplet is moved to a central electrode and **C)** is divided into two even droplets of 150 nl. **D–E)** The small droplets are transportated to electrodes at the right side of the structure, respectively. **F–H)** The droplets

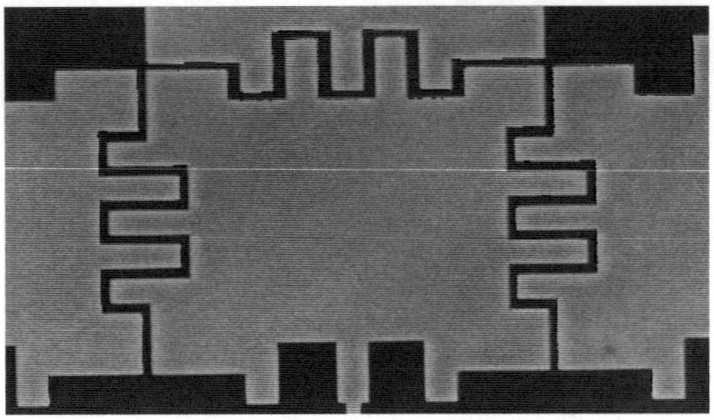

Figure 3: Interferometry measurement of an EWOD electrode (1.4 mm × 1.4 mm). 20 nm chromium and 35 nm platinum are sputtered onto a quartz substrate. The surface is covered by an aluminum oxide layer of 600 nm thickness.

are remerged and moved back to the reservoir. The same operation can be shown with DI-water. In this case the spherical droplet shape is less deformed while the droplet displacement from electrode to electrode is more precise. This is due to a higher surface tension of DI-water compared to DNA solutions with additional TWEEN® 20 which significantly lowers the surface tension.

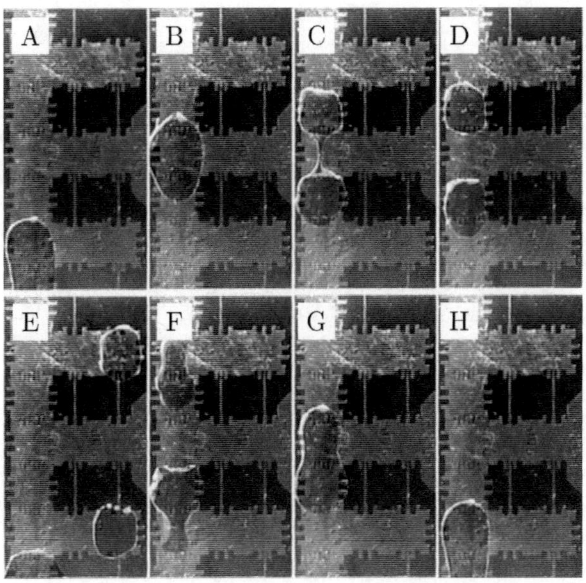

Figure 4: Transport of DNA solution using EWOD. The created droplet has a volume of approximately 300 nlbefore the splitting and afterwards a volume of ca. 150 nl. The droplet volume consists of 10 ng/µl DNA solution in water with 0.01 wt% TWEEN® 20 added. Electrical parameters: $U = 80\,\mathrm{V_{pp}}$, $f = 250\,\mathrm{Hz}$ square wave, switching time $t = 500\,\mathrm{ms}$.

It should be noted that DI-water droplets can be switched more rapidly (250 ms) without any errors. Moreover, an electrowetting system driven with pure water is less prone to dielectric breakdown because of the lower conductivity of DI-water compared to electrolyte solutions. The EWOD system can be driven for several hours at voltages from $80\,\mathrm{V_{pp}}$ to $100\,\mathrm{V_{pp}}$. At higher voltages defects at the electrode edges occure, which finally lead to a dielectric breakdown. This is due to a higher density of the electric field lines at the edges. Good repeatability of the droplet movement indicates that the addition of 0.01 wt% TWEEN® 20 to the solution minimizes the adsorption of nucleic acids onto the teflon layer.

3.2. Combination of EWOD device and microscopy

In a first experimental setup of the assembled liquid handling system and the confocal microscope two droplets (DI-water and Alexa488 solution) have been merged in the focus of the confocal

IOP Conf. Series: Journal of Physics: Conf. Series **922** (2017) 012005 doi:10.1088/1742-6596/922/1/012005

microscope using EWOD technique. The measured fluorescence intensity over time is shown in figure 5. After a short lag time of about 3 min, a sigmoidal increase of the fluorescence intensity until minute 12 can be seen (data points), after which a linear signal increas can be observed. The results indicate the diffusion of Alexa488 molecules from the highly concentrated droplet volume to the low concentrated dropled volume until an equilibrium is reached. The linear increase of the intensity is due to evaporation. Thus leading to an observable increase of concentration over time. The dotted line displays the corrected intensity and shows a plausible curvature for free molecule diffusion. For the modeling of the evaporation a cylindric geometry of the merged droplet is assumed.

At first glance the large time scale of the measurement surprises. But it has to be considered

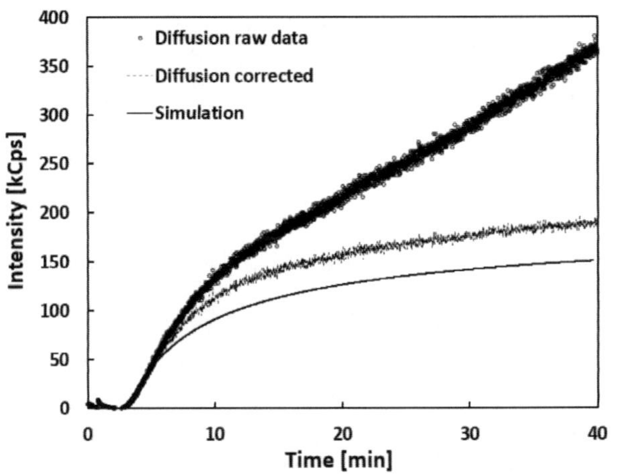

Figure 5: Diffusion of Alexa 488 after merging of two droplets. Fluorescence measurement with confocal microscope using 63x oil emulsion objective.

that after merging almost exclusively diffusion is responsible for molecule transport within the droplet volume over a distance of about 0.5 mm. For comparision simulation data for a solely diffusion-driven mixing of droplet volumes is added to the graph. The diffusion is modeled as two volumina with infinite extension. At the very beginning the molecule distribution at the

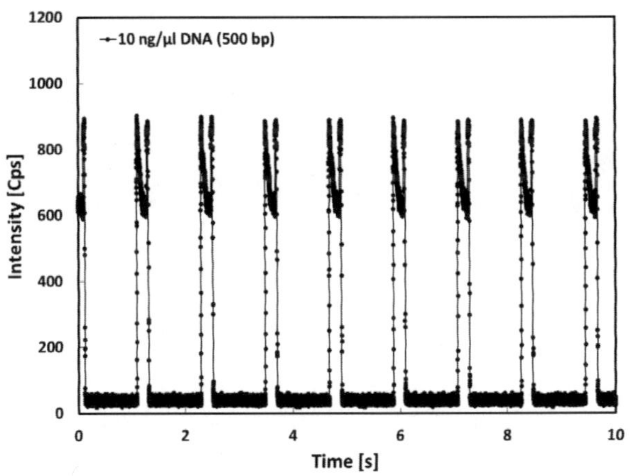

Figure 6: Movement of DNA solution in and out of the detection volume of the microscope. 10 ng/µl DNA solution with 0.01 wt% TWEEN® 20 added, 10x objective

transition line from one volume to the other is like a step function. From now on the mixing of the two volumes is only driven by diffusion. As it can seen the time scales match the previous data curves. The discrepancy between measurement and simulation can be explained by a small convective mixing term in the droplet experiment which was not considered in the model of the

molecule diffusion. Another uncertainty is related to the evaporation effect and its modelling. The evaporation can be supressed by using silicone oil or dodecane as ambient fluid.
The intensity measurements during several movement sequences of DNA solution in and out of the detection volume of the confocal microscopy can be seen in figure 6. The intensity signal sharply increases from 50 Cps to 900 Cps each time the droplet is transported into the detection volume. Another peak is found whenever the droplet is moved out of the detection spot. Between these two sharp peaks an intensity decrease from 900 Cps to 600 Cps is observable. The signal intensity of 50 Cps corresponds to the background noise of the measurement in the EWOD device. The intensity drop from 900 Cps to 600 Cps between the droplet movements is presumably due to bleaching of the fluorescence dye. Because of the highly reproducible intensity curves for each droplet movement a good mixing of the droplet volume can be assumed. The excellent reproducibility of the peaks allows the derivation of underlaying models to describe the bleaching of the fluorescence molecules. It can be concluded that the experiments prove the feasibility of time-resolved measurements in droplets after transporting via electrowetting.

4. Conclusion

In this work an EWOD based liquid handling technique has been developed and combined with a confocal microscope. The developed electrowetting system for liquid handling can manipulate DI-water droplets and DNA solutions of 10 ng/µl concentration repeatedly and reliably. Additionally, standard operating modes like droplet transport, droplet merging, droplet splitting and taking droplets from a reservoir has been shown. Thus, a powerful liquid handling system has been established. The proof-of-concept i.e. the combination of an EWOD device with a confocal microscope was displayed by the merging of two droplets and the measurement of fluorescence increase due to molecule diffusion from one droplet to another. In addition, a small droplet of DNA solution was moved in and out of the detection spot of the microscope. The measured intensity signals are reproducible without any signal loss from peak to peak. This indicates the good suitability of the combined system for further molecule interaction measurements.
In this work the potential of EWOD based liquid handling techniques combined with high resolution confocal microscopy has been shown. Further experiments are planned to investigate the interaction of nucleic acids on a single molecule level. This knowledge may help to profoundly increase the understanding in biological mechanisms and interactions. The identification of new drug candidates and the development of new therapeutics will benefit from these insights.

5. References

[1] L. Mandel. *Proceedings of the Physical Society*, 72:1037–1048, 1958.
[2] B. Greiner, S. Krueger, M. Droege, B. Steinbock, G. Aragon, A. Linnemann, and H. Mathis. *Biophotonik*, 1:24–27, 2016.
[3] T. Hirschfeld. *Applied Optics*, 15:2965–2966, 1976.
[4] G. Lippmann. *Annales de Chimie et de Physique*, 5:494–549, 1875.
[5] M. G. Pollack and R. B. Fair. *Applied Physics Letters*, 77(11):1725–1726, 2000.
[6] H. Moon, S. K. Cho, R. L. Garrell, and C-J. Kim. *Journal of Applied Physics*, 92(7):4080–4087, 2002.
[7] S. K. Cho, H. Moon, and C-J. Kim. *Journal of Microelectromechanocal Systems*, 12(1):70–80, 2003.
[8] V. N. Luk, G. CH. Mo, and A. R. Wheeler. *Langmuir*, 24:6382–6389, 2008.
[9] A. Rival, D. Jary, C. Delattre, Y. Fouillet, G. Castellan, A. Bellemin-Compte, and X. Gidrol. *Lab on a Chip*, 59:327–333, 2014.
[10] H-Y. Huang, H-H. Shen, C-H. Tien, C-J. Li, S-K. Fan, C-H. Liu, W-S. Hsu, and D-J. Yao. *PLoS*, 10(5):1–15, 2015.
[11] N. Avci, P. F. Smet, J. Lauwaert, H. Vrielinck, and D. Poelman. *Journal of Sol-Gel Science and Technology*, 24:6382–6389, 2011.
[12] H. Verheijen and M. Prins. *Review of Scientific Instruments Journal*, 15:6616–6620, 1999.
[13] N. Verplanck, Y. Coffinier, V. Thomy, and R. Boukherroub. *Nanoscale Research Letters*, 2(12):577–596, 2007.

Nonlinear numerical analysis and experimental testing for an electrothermal SU-8 microgripper with reduced out-of-plane displacement

Rodica-Cristina Voicu[1], Muaiyd Al Zandi[2], Raluca Müller[1], Changhai Wang[2]

[1] National Institute for Research and Development in Microtechnologies-IMT Bucharest, Erou Iancu Nicolae str., 126A, 077190, Romania

[2] Institute of Sensors, Signals and Systems, School of Engineering & Physical Sciences, Heriot-Watt University, Edinburgh EH14 4AS, UK

rodica.voicu@imt.ro

Abstract. This paper reports the results of numerical nonlinear electro-thermo-mechanical analysis and experimental testing of a polymeric microgripper designed using electrothermal actuators. The simulation work was carried out using a finite element method (FEM) and a commercial software (Coventorware 2014). The biocompatible SU-8 polymer was used as structural material for the fabrication of the microgripper. The metallic micro-heater was encapsulated in the polymeric actuation structures of the microgripper to reduce the undesirable out-of-plane displacement of the microgripper tips, and to electrically isolate the micro-heater, and to reduce the mechanical stress as well as to improve the thermal efficiency. The electro-thermo-mechanical analysis of the actuator considers the nonlinear temperature-dependent properties of the SU-8 polymer and the gold thin film layers used for the micro-heater fabrication. An optical characterisation of the microgripper based on an image tracking approach shows the thermal response and the good repeatability. The average deflection is ~11 μm for an actuation current of ~17 mA. The experimentally obtained tip deflection and the heater temperature at different currents are both shown to be in good agreement with the nonlinear electro-thermo-mechanical simulation results. Finally, we demonstrate the capability of the microgripper by capture and manipulation of cotton fibres.

1. Introduction

Micromanipulators developed as micro-robotic arms or as microgrippers are important tools for handling, manipulating, picking and placing delicate micro-objects. The applications of such devices include biological micro-manipulation (living cells, blood vessels, and tissues), and micro-assembly of Micro-electro-mechanical Systems MEMS/MOEMS components (micro-gears, micro-mirrors, lenses, fibers) and micro-robotics applications.

Usually, such microgrippers are often integrated with actuators that are used for energy conversion, motion generation and force production [1-3]. Different actuation techniques can be integrated with MEMS in order to obtain the desired performance, such as electrostatic, piezoelectric, thermal, shape memory, pneumatic or magnetic strategies. The advantage of the thermal actuators mainly depends on the coefficient of thermal expansion (CTE) of the structural material. Frequently, the structural material of the actuators is silicon, polysilicon or aluminum. But, for example, due to the small CTE, silicon produces small mechanical deflection. The alternative materials used for the structural part of the actuator are polymers which are preferred for thermal actuation due to the low Young modulus and high CTE [4-5].

Different microgrippers have been considered using the SU-8 polymer based thermal actuators designed on different configurations such as, U-shape or V-shape based designs. This is proving interest in the bio-micro-manipulation domain [2-14]. SU-8 is a highly cross-linked epoxy-photo-polymer which has been the preferred polymer material for fabrication of biocompatible devices. The SU-8 polymer has a relatively large coefficient of thermal expansion (CTE) of 52 ppm/°C [15], good mechanical strength with a Young's modulus of between 2-5 GPa [15] and a good thermal stability with a glass transition temperature of 210°C [15] which make it a proper material for manufacture of thermal actuators.

The thermo-mechanical properties of the polymers, such as SU-8, and the electrical conductivity of the metals used for micro-heaters are essential for the behavior of the thermal actuators. Previous research has studied the nonlinear mechanical and thermal properties of fully cross-linked SU-8 in order to be used for modeling of thermal actuators. The effect of the temperature on the Young's modulus, Poisson ratio and CTE was reported in different studies [4-5].

In this work, we present the results of numerical nonlinear electro-thermo-mechanical analysis and testing of an SU-8 polymeric microgripper designed using the electrothermal actuators. The SU-8 microgripper can be utilized for bio-manipulation, assembly and micro-robotic applications. In order to confirm the performance of the microgripper, the actuator was investigated by numerical simulation based on a finite element method (FEM) and the Coventorware (2014). The biocompatible SU-8 polymer was used in the fabrication process as the structural material. The gold metallic micro-heater was encapsulated in the polymeric actuation structures of the microgripper to reduce the undesirable out-of-plane displacement of the microgripper tips, to electrically isolate the micro-heater, to reduce the mechanical stress and to improve the thermal efficiency. The electro-thermo-mechanical analysis of the actuator included the nonlinear temperature-dependent properties of the SU-8 polymer and also of the gold thin metal layer used for the micro-heater fabrication. The characterization of the gripper displacement was based on an image tracking approach. The experimentally obtained tip deflection and the heater temperature at different currents are both shown to be in good agreement with the nonlinear electro-thermo-mechanical simulation results. Finally, we demonstrate the capability of the microgripper by manipulation of cotton micro-fibers.

2. Design and Fabrication

The design and the fabrication of the polymer microgripper using a sandwich structure with the metallic micro-heater embedded in the polymeric actuation structures of the microgripper (figures 1 and 2) were reported before [11-14]. The fabrication is based on the structure release using the designed chips with different microgripper structures (figure 3).

2.1 Design

The polymer microgripper was designed using the thermal actuators in normally close operation. The initial opening of the microgripper is of 10 μm (figure 2). When the gripper is electro-thermally actuated, the tips will open and can grip a micro-object. The total length of the gripper is 1300 μm. The arms were designed with a width of 20 μm. A metallic micro-heater is embedded between two SU-8 layers with the same thickness of around 10 μm. The track width of the heater was designed to be 10 μm and the thicknesses of the heater layers were 10/300/10 nm for Cr/Au/Cr films.

The microgripper was designed symmetrically with embedded metallic micro-heaters in the structural material of the grippers (figure 1), the SU-8 polymer, in order to reduce the undesirable out-of-plane displacement of the gripper, to obtain the electrical isolation of the heaters and to reduce the mechanical stress that can occur in the structure [12-14].

The substrate where the structure will be fixed after fabrication can be either a silicon or a glass wafer.

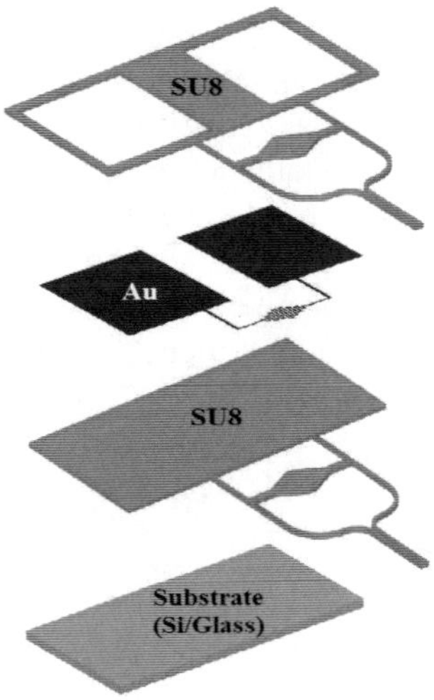

Figure 1. Exploded view of the 3D microgripper model with embedded heater between two SU-8 layers.

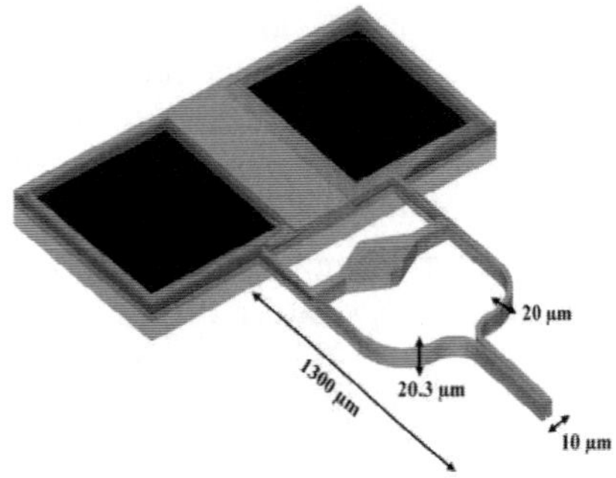

Figure 2. 3D model of the microgripper used in FEM simulations (Coventorware).

2.2 Fabrication

The optimized gripper design planned for manufacture consists of a symmetrical structure of three material layers with the metallic layer for the heater embedded between two SU-8 based structure layers having the same thickness, as described previously [13-14].

The fabrication of the microgripper is based on a 3-mask process. The two masks used for the SU-8 structural gripper configuration have similar designs, one of the masks having openings for producing access windows on the metallic pads. The third mask was used for the micro-heater configuration. The heater and pads were obtained using a lift-off process using an AZ photoresist consisting of metal films of Cr/Au/Cr. The OmniCoat stripper (MicroChem) was used in order to completely release the final structures. The details of the fabrication work have been reported also previously in [13]. Figure 3 shows that the fabricated microgripper before release and the released chip with different SU-8 microgrippers are well configured.

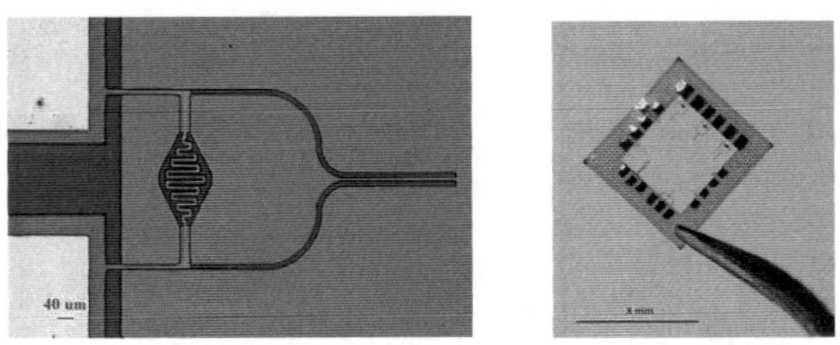

a) b)

Figure 3. Optical microscope pictures: a) a fabricated SU-8 microgripper before release; b) one chip with released SU-8 microgrippers;

3. Numerical Simulation and Experimental Testing

Numerical analysis, characterisation and experimental testing were performed in order to determine the microgripper behaviour in air. Each microgripper was mounted on a hexapod based robotic system. A microscope based optical vision system and an image tracking software was used to realise the measurements regarding the displacement of the gripper tips. Further, the measured TCR was used to determine the heater temperature in gripper operation [13-14].

IOP Conf. Series: Journal of Physics: Conf. Series **922** (2017) 012006 doi:10.1088/1742-6596/922/1/012006

3.1 First Electro-thermo-mechanical Analysis

Coupled electro-thermo-mechanical FEM simulations were completed using the MemMech simulator (Coventorware). A simplified 3D microgripper model (figure 2) was meshed using hexahedral parabolic elements (Extruded bricks). The number of volume elements was optimized by choosing the proper size and using the Split and Merge algorithm. The surface boundary conditions were set for the simulations. The initial temperature of the whole structure and the temperature of the environment were considered to be T_0=27 °C, with respect to the Coventorware settings for this type of analysis. The effect of radiation was neglected since the polymeric structure works at low operating temperatures.

For the first simulation analysis, constant values for the material properties of SU-8 were used. The Young's modulus of the SU-8 was set at 4.6 GPa. The CTE was fixed at 52 ppm/°C and the thermal conductivity at $2x10^5$ pW/μmK. For the gold layer, we used a Young's modulus of 77 GPa reported for thin films. For the thin Cr/Au/Cr films, a measured temperature coefficient of resistance (TCR) of 0.00147/°C was used [13-14] in order to take into account temperature effects on electrical properties. The electrical conductivity of the gold layer, σ was set using equation (1):

$$\sigma(T) = 1/\{\rho_0\,[1 + \varepsilon \cdot (T - T_0)]\} \quad (1)$$

where ρ_0 is the resistivity at T_0, ε is the TCR of the Cr/Au/Cr layer. The air convection coefficient was variated between 150 [1] and 500 W/m^{2}K in order to obtain the best fit with the experimental results.

The distribution of the temperature and mechanical behavior of the microgripper at 15mA actuation current can be observed in figure 4. The temperature remain around 27 °C at the tips and the out-of-plane displacement under 0.1 μm. The simulated values of the temperatures reached in the arms are presented in figure 5 a) and the simulated in-plane deflections of the tips as function of electrical current in figure 5 b) when the convection coefficient was varied. When the convection coefficient is near 300 W/m^{2}K the results of simulation and measurements are in good agreement regarding the temperatures (figure 5 a)) and electrical currents (figure 5 c)). The results for displacements are not in such good agreement (figure 5 b)).

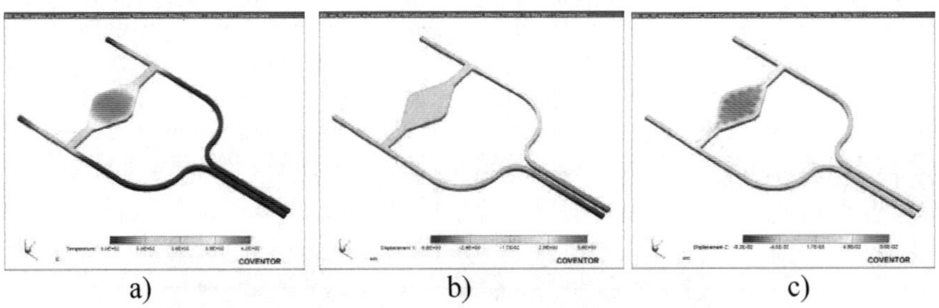

a) b) c)

Figure 4. FEM coupled electro-thermo-mechanical simulation results at 15 mA: a) Temperature distribution in the microgripper; b) In-plane deflection; c) Out-of-plane deflection (Coventorware).

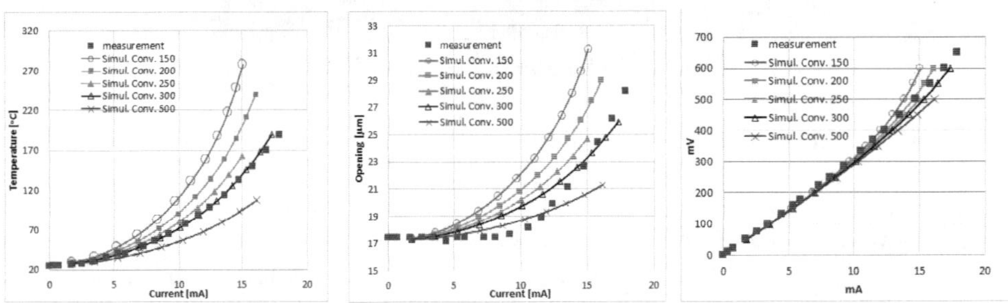

a) b) c)

Figure 5. Simulation results vs measurements: a) Temperature values as electrical current; b) Tips opening vs electrical current; c) Electrical Current vs potential.

3.2 Second Electro-thermo-mechanical Simulations

In order to improve the agreement between the simulation results regarding the displacements and the measurements results, the mechanical and the thermal material properties of the SU-8 were varied as a function of temperature. The dependence of Young's modulus on temperature was modelled (2) using

$$E(T) = E_0 \exp(-T/T_C) \qquad (2)$$

where, the constant E_0 is 5.76 GPa and the constant temperature T_C is 55.82 °C [4-5]. Similarly, the Poisson's ratio with temperature was modelled (3) as

$$\nu(T) = \nu_0 + \nu_1 T + \nu_2 T^2 \qquad (3)$$

where, the constants, ν_0, ν_1 and ν_2 are obtained from the experimental results [5]. The small nonlinearity in CTE for the SU-8 as function of temperature was modelled (4) as

$$CTE(T) = CTE(T_0)[1 + \varepsilon_{CTE} \cdot (T - T_0)] \qquad (4)$$

where, the constant $CTE(T_0)$ is the CTE at ambient temperature and $\varepsilon_{CTE} = 0.00175\ ^0C^{-1}$ [4].

All simulations were performed for a convection coefficient of 300 W/m^{2}K. The CTE value was varied from 52 to 30 ppm/°C for SU-8. The best agreement was obtained for a CTE of 30 ppm/°C (figure 6). The out-of-plane displacement remains under 0.1 μm in each actuation stage. Further investigations can be completed varying the SU-8 material properties, such as Young's modulus, Poisson ratio and the CTE which strongly influence the mechanical behavior of the gripper (figure 6 b)).

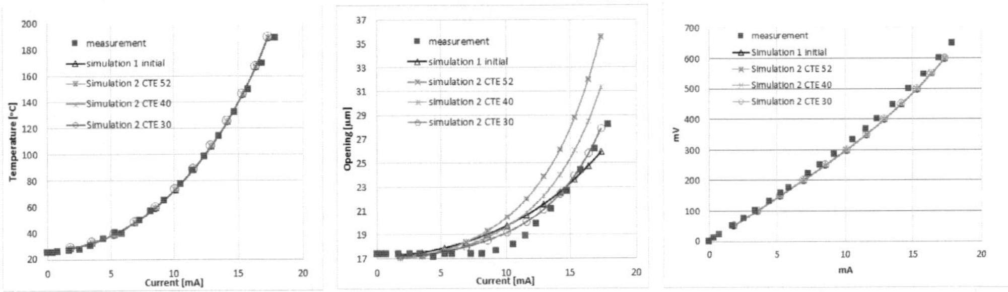

a) b) c)

Figure 6. Simulation results vs measurements: a) temperatures as function of actuation current; b) opening of the microgripper tips vs current; c) current vs potential.

3.3 Experimental Testing

The experimental testing was performed in air to analyze the openings of the microgripper when it is actuated (figure 7 a) and b)). In order to demonstrate the microgripper performance, different tests were performed by manipulating cotton fibers in air using the microgripper mounted on a robotic system (figure 7 c) and d)).

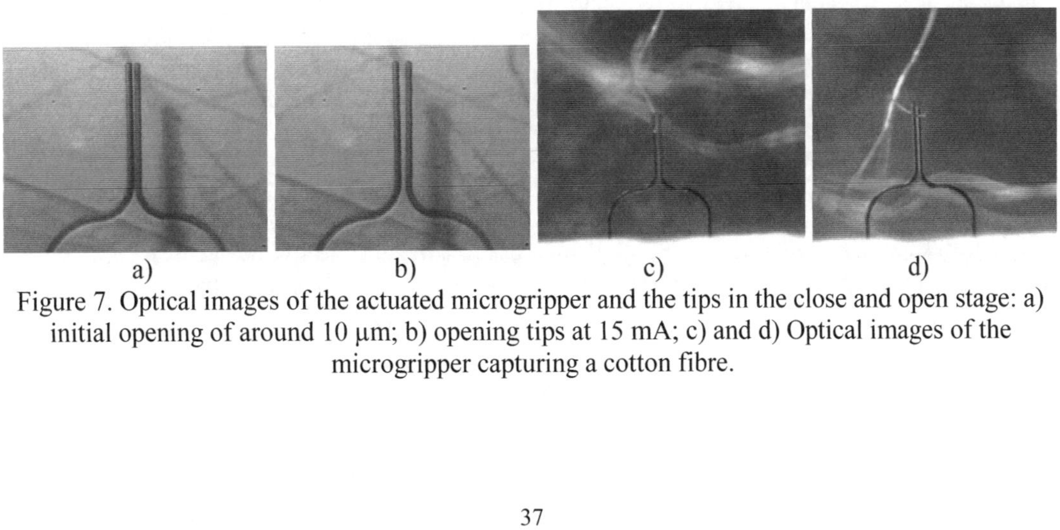

a) b) c) d)

Figure 7. Optical images of the actuated microgripper and the tips in the close and open stage: a) initial opening of around 10 μm; b) opening tips at 15 mA; c) and d) Optical images of the microgripper capturing a cotton fibre.

4. Conclusions

Electro-thermo-mechanical analysis and experimental testing of a polymeric microgripper based on electrothermal actuation have been presented. The simulations of the actuator including nonlinear temperature-dependent properties of the SU-8 polymer and the gold thin metal layer used for the micro-heater fabrication demonstrates a good agreement between the results of experiments and numerical analysis. The average in-plane displacement is ~11 μm for an actuation current of ~17 mA. The out-of-plane displacements remain under 100 nm. We demonstrate the gripping capability of the microgripper by capture and manipulation of cotton fibers.

Acknowledgments

This work was supported by a grant of the Romanian National Authority for Scientific Research and Innovation, CCCDI – UEFISCDI, project number 22/2016, within PNCDI III, ERA-MANUNET-II-Robogrip project "Microgrippers as end-effectors with integrated sensors for microrobotic applications (ROBOGRIP)" (2016-2017).

References

[1] Chu J, Zhang R, Chen Z 2011 A novel SU-8 electrothermal microgripper based on the type synthesis of the kinematic chain method and the stiffness matrix method *J. Micromech. Microeng.* **21** 5, pp. 054030

[2] Jia Y and Xu Q 2013 MEMS microgripper actuators and sensors: The state-of-the-art survey *Recent Patents on Mechanical Engineering* **6** 2

[3] Que L et al 2001 Bent-beam electrothermal actuators. Part I: Single beam and cascaded devices *J. Microelectromech. Syst.* **10** pp. 247–254

[4] Viannie L R et al 2016 Fabrication and nonlinear thermomechanical analysis of SU8 thermal actuator *J. of Microelectromech. S.* **25** 1 pp. 125-133

[5] Chung S, Park S 2013 Effects of the temperature on mechnaical properties of SU-8 photoresist material *J. Mech.Sci. Technol.* **27** 9 pp. 2701-2707

[6] Colinjivadi K S, Lee J-B, Draper R 2008 Viable cell handling with high aspect ratio polymer chopstick gripper mounted on a precision manipulator *Microsyst. Technol.* **14** pp. 1627-1633

[7] Solano B 2008 A microgripper for single cell manipulation, *PhD thesis, Durham University*

[8] Zhang R et al 2012 A multipurpose electrothermal microgripper for biological micro-manipulation *Microsyst. Technol.* **19** 1 pp. 89-97

[9] Kim K et al 2004 Metallic microgripper with SU-8 adaptor as end-effectors for heterogeneous micro/nano assembly applications *Microsyst. Technol.* **10** 10 pp. 689-693

[10] Wang Z, Shen X and Chen X 2015 Design, modeling, and characterization of a MEMS electrothermal microgripper *Microsyst. Technol.* **21** 11 pp. 2307-2314

[11] Voicu R et al 2014 Design and Analyses of a laterally driven MEMS electro-thermally microgripper, in *Proc. 8-th Int. Conference & Exhibition on Integration Issues of Miniaturized Systems - MEMS, NEMS, ICs and Electronic Components (SSI)* pp. 591-594

[12] Voicu R and Müller R 2013 New electro-thermally actuated micromanipulator with optimized design and FEM simulations analyses *Proc. Conf. Design, Test, Integration & Packaging of MEMS/MOEMS (DTIP)* pp. 1-6

[13] Al-Zandi M H, Voicu R-C, Muller R, Wang C 2016 Testing and Characterisation of Electrothermal Microgrippers with Embedded Microheaters, *Proc. Conf. Design, Test, Integration & Packaging of MEMS/MOEMS (DTIP),* pp. 256-260

[14] Al-Zandi M H, Wang C, Voicu R-C, Müller R 2017 Measurement and Characterisation of Displacement and Temperature of Polymer Based Electrothermal Microgrippers, *Microsyst. Technol.* DOI 10.1007/s00542-017-3298-8

[15] MicroChem 2016 SU-8 2000 Permanent Epoxy Negative Photoresist [Online] http://www.microchem.com/

IOP Conf. Series: Journal of Physics: Conf. Series **922** (2017) 012007 doi:10.1088/1742-6596/922/1/012007

Smart design piezoelectric energy harvester with self-tuning

L G H Staaf[1], E Köhler[1], P D Folkow[2], P Enoksson[1]

[1] Department of Microtechnology and Nanoscience, Chalmers University of Technology, Gothenburg, Sweden

[2] Division Dyanmics, Chalmers University of Technology, Gothenburg, Sweden

V96staaf@chalmers.se

Abstract. Piezoelectric energy harvesting on a gas turbine implies constraints like high temperature tolerance, size limitation and a particular range of vibrations to utilise. In order to be able to operate under these conditions a harvester needs to be space effective and efficient and to respond to the appropriate range of frequencies. We present the design, simulation and measurements for a clamped-clamped coupled piezoelectric harvester with a free-sliding weight, which adds self-tuning for improved response within the range of vibrations from the gas turbine. We show a peak open circuit voltage of 11.7 V and a 3 dB bandwidth of 12 Hz.

1. Introduction
The coming of the internet of things requires that we now develop alternative energy sources to replace or support today's and tomorrow's batteries. Alternative energy sources such as energy harvesters have been focused on by many researchers during the past years [1]. Energy harvesters convert ambient energy in our surrounding, like solar irradiation, wind, heat and mechanical vibrations into electric energy. For a gas turbine the main ambient energies are heat and mechanical vibrations. The thermal energy is concentrated to certain areas within the gas turbine, while vibrations are more or less available everywhere, making piezoelectric energy harvesting a viable option. The main challenge for piezoelectric energy harvesting is to maintain a sufficient power output over a broad bandwidth; much research has been done on broadening the harvesting bandwidth [1-2]

In this paper we report on the design, simulation and measurements of a piezoelectric harvester with self-tuning for wider bandwidth and coupled piezoelectric cantilevers to maintain a high power output by extended strain distribution. The harvester is designed to meet the gas turbines specific conditions on size, temperature and frequency. Previous work [3] has shown that by utilizing a distributed stress pattern over the whole cantilever, the complete area of the piezoelectric cantilever is used and not only the clamped end, thereby yielding a higher power output. In previous work [4] this distributed stress pattern was observed for a cantilever with one end fixed and the other end coupled to a second (top) cantilever (which showed a stress pattern similar to a single cantilever with one end fixed). In the present design both cantilevers are clamped and coupled at each end, hence using all available piezoelectric capacity to optimize the power output. To extend the bandwidth, passive self-tuning is introduced by a free sliding weight. In previous work a free sliding proof mass has been used on a thin fixed-fixed beam. When the proof mass slides to one end of the thin beam it can get stuck there even if

IOP Conf. Series: Journal of Physics: Conf. Series **922** (2017) 012007 doi:10.1088/1742-6596/922/1/012007

the frequency is changed [5-6]. To avoid this problem we have made our middle beam thicker, in order to lower or remove the impact of the natural frequencies that might force the weight to one end. The movement range of the proof mass has also been limited to prevent the occurrence of it getting stuck at one end.

Simulation

The harvester was modelled in the simulation tool COMSOL. Figure 1 shows a schematic image of the harvester; it consists of two piezoelectric cantilevers that are clamped to an attachment at one end. On the other end the cantilevers are attached to a coupling that connects the two cantilevers by an aluminium beam where a sliding weight is placed. The targeted frequency range is 370 - 380 Hz with self-tuning from the weight. The simulated harvester is tuned to 380 Hz, with a fixed weight at the centre of the middle beam, and the simulation provides the required length of the cantilevers on the real harvester. In figure 2 the mode shape of the resonance frequency mode at 380 Hz is shown. From experiment when the harvester is at the lowest resonance frequency of 374 Hz, the weight is at the outer end of the middle beam according to measurement observation. The correlating mode is simulated and presented in figure 4 and shows that both cantilevers are utilized when the harvester is in its first natural frequency. The idealized COMSOL model has slightly higher natural resonance frequencies than the measured ones. The weight placed on the middle beam is imitating the real sliding weight but in COMSOL statically placed at different locations (0, 7, 10 mm) from the center of the middle beam. From experimental result, we can see that the weight is moving away from the center and then slides back towards the center when being out of the frequency.

Looking at the simulated open voltage output for the different weight positions, figure 5, a sliding weight that is further from the center (green curve) gives the system a lower first resonance frequency than for the other two positions (red and blue curve). The total bandwidth will be a total curve with the outer left and outer right boundary of the three simulated output curves, since we know from experiment that the weight is shifting its position due to the applied driving frequency. The conclusion from COMSOL simulation is that the sliding weight placement has a clear correlation with the natural resonance frequency of the harvester.

2. Experimental setup

The characterization of the harvester is done on a sinusoidal excited shaker table. The cantilevers electrodes are connected to multimeters measuring the open circuit voltage over the frequency span 330-410 Hz. In figure 4 the harvester assembly is shown. The front wall is removed for a better view from the side. In case 1 (C1), figure 4, the cantilevers are 24 mm long and the length of the middle beam is 16 mm. The coupling weight is in total 3.87 g, where the sliding weight is 0.87 g. The active middle beam is 1 mm thick, 3 mm broad and 16 mm long. The natural frequency of the middle beam is 2974 Hz, which is far above our targeted frequency range of 370 – 380 Hz.

In case 2 (c2), figure 5, the top and bottom cantilever lengths are 26 mm and 28 mm respectively. The middle beam is 1 mm thick, 3 mm wide and 27.5 mm long. The natural frequency of the middle beam is 1006 Hz, still above our targeted frequency range of 370 – 380 Hz.

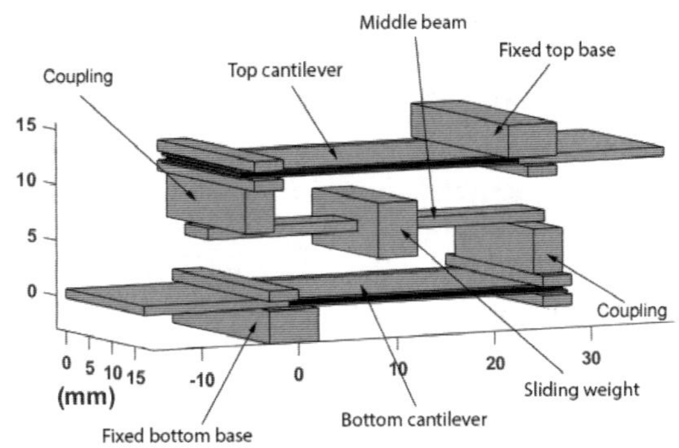

Figure 1. Schematic image of the harvester setup in COMSOL.

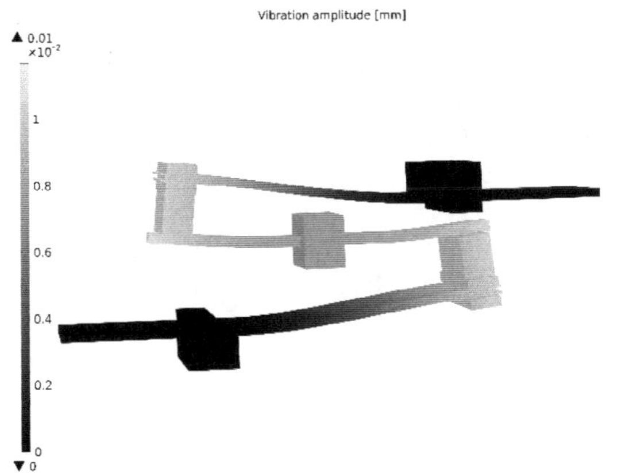

Figure 2. The mode shape at resonance frequency at 380 Hz from simulation

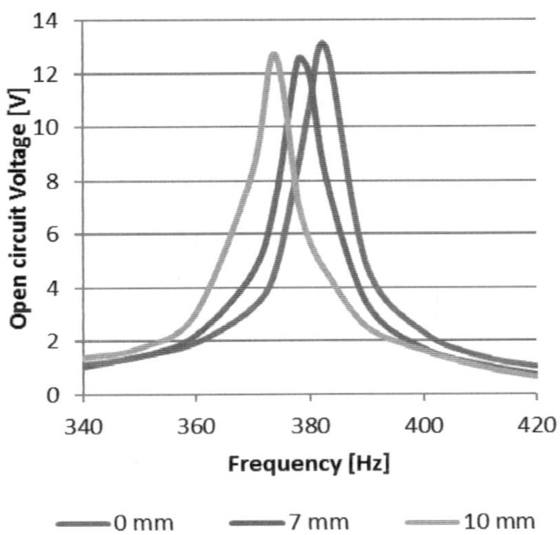

Figure 3. Weight sweep spectrum output with the weights position given from center: [0, 7, 10] mm

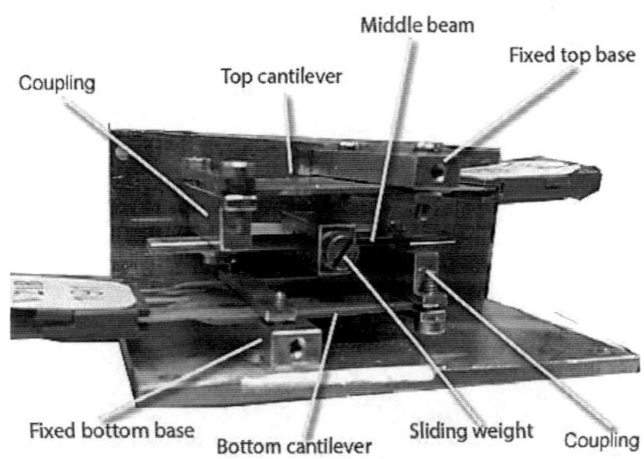

Figure 4. The setup of the harvester, at measurement 1, tuned to 373 Hz

3. Result

Table 1. Comparison of C1 and C2

	Primary mode (Hz)	*Open circuit voltage (V)*	3 dB open circuit voltage (V)	*3 dB bandwidth (Hz)*
C1 16 mm beam, sliding weight	373	3.12	2.2	12
C1 16 mm beam, fixed weight	378	3.26	2.3	8
C2 27.5 mm beam, loose sqrew	373	11.2	7.9	10
C2 27.5 mm beam, extremely loose screw	372	11.71	8.28	12
C2 27.5 mm beam, fixed weight	378	11.5	8.13	8

IOP Conf. Series: Journal of Physics: Conf. Series **922** (2017) 012007 doi:10.1088/1742-6596/922/1/012007

3.1 Case 1

The first measurement series was made with 370 - 380 Hz as target frequency. The setup was manually tuned to a resonance frequency of 373 Hz. Even though the movement of the proof mass was visibly much more limited than expected, a small effect could be seen when comparing to a completely fixed weight. In figure 6 the measured result (M1) for the total output with non-fixed and fixed weight with a 16 mm long middle beam is shown for the primary frequency mode of 373 Hz with a 3 dB bandwidth of 12 Hz. The measured open circuit voltage output peaks at 3.1 V and 2.2 V at the 3 dB border. The self-tuned harvester has nearly as high voltage output as the harvester with the fixed weight, but shows a near 50% increase in bandwidth as shown in table 1.

3.2 Case 2

From the first measurement we could conclude that the non-fixed weight can increase the bandwidth compared with a fixed weight. Shorter cantilevers are harder to tune than longer, therefore new simulations were performed with longer cantilevers. For the second measurement three parameters were changed, two based on simulation results and one mechanical (a more reliable weight fixture). The middle beam was set to be as long as possible (27.5 mm) due to the current setup geometric constraints as shown in figure 5. In this setup the top cantilever is 26 mm long and the bottom one is 28 mm long. The target frequency was still 370 - 380 Hz; with longer cantilevers and longer middle beam the harvester showed resonance at 372 Hz.

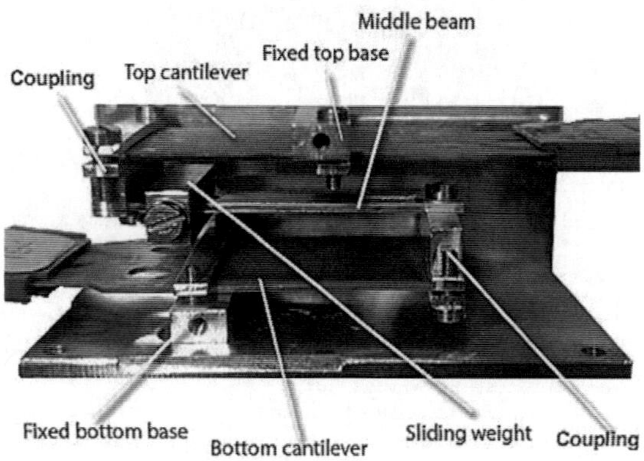

Figure 5. C2 setup with 27.5 mm middle beam and longer top (26 mm) and bottom (28 mm) cantilevers compared with measurement 1 setup.

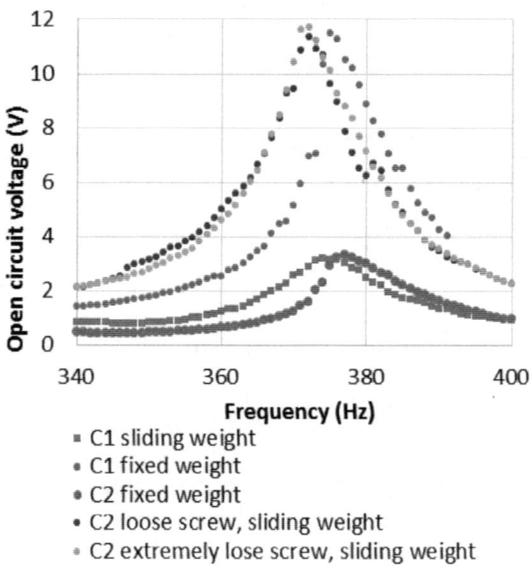

Figure 6. C2 open circuit voltage output with different screw tightness and compared with C1 which had a 16 mm middle beam and 3.5 times lower output.

In the second measurement the new weight fixture screw was tested for two different settings; *loose* (-½ screw turn from fixed condition) and extremely loose (unscrewed). For both settings, the weight could be seen to move to different positions when the frequency was changed. By slowly increasing the frequency we could see that after 2-3 seconds at a specific frequency, the weight moved to another position, where it henceforth remained if the frequency was kept constant. At the start of a measurement sweep the weight was positioned at the middle of the beam. When the frequency reached 357 Hz it moved to the left end (refer to figure 5) and at 368 Hz it started to slide back and forth up to 372 Hz from when it remained stable at the left side. When the frequency sweep was downwards the sliding weight positioned itself accordingly at the same frequency as on the sweep upwards. In figure 6 the open voltage output for the screw settings are compared; it is inferred that a loose screw setting seems preferable.

Compared to C1 we can conclude that C2 had approximately 3.5 times higher open circuit voltage output of 11.7 V (table 1) for the same volume, which might occur from that the natural frequency of the middle beam which is more close to the resonance frequency of the harvester C2 than in C1. Hence with a lower natural frequency for the middle beam a higher impact on the open circuit voltage might occur. Thinner middle beams with lower natural frequency have to be tested for further evidence of this theory. The bandwidth was the same 3 dB bandwidth range of 12 Hz C1 and C2

4. Conclusion

Our second generation of coupled harvester displayed a bandwidth of 12 Hz and an open circuit voltage of 11.7 V, which is 3.5 times higher, with the same volume, than our first harvester implementation. The higher open circuit voltage might occur from that the natural frequency of the middle beam is closer to the resonance frequency of the harvester on the second measurement. Future harvester improvements are foreseen when tuning 1) the proof mass weight, 2) the stiffness of the couplings and 3) the thickness of the middle beam. Despite improvements remaining to be implemented, the tested harvester demonstrates the bandwidth broadening impact of a passively sliding weight and high open circuit voltage which looks promising for our gas turbine application.

Acknowledgment
The European project STARGATE, SMART Memphis and Vinnova UDI 1 energy harvesting toolkit are greatly acknowledged for their financial support.

References

[1] Zhu D, Tudor M J and Beeby S P, 2009, *Meas. Sci. Technol. 21*
[2] Wu H, Tang L, Yang Y and Soh C K 2014, *J. of Int. Mat. Sys. and Struct, vol 25 1875-1889*
[3] Staaf L G H, Köhler E, Parthasarathy D, Lundgren P and Enoksson P, 2015, *Sensors and Actuators A: Physical Volume 229, p 136–140*
[4] Staaf L G H, Köhler E, Lundgren P and Enoksson P, 2014, ***PowerMEMS 2014** (1742-6588). Vol. 557*
[5] Gregg C G, Pillatsch P and Wright P K, 2013, *J. of Phys.Conference Series 557 (2014) 012123*
[6] Pillatsch P, Miller L M, Halvorsen E, Wright P K, Yeteman E M and Holmes A S, 2013, *J. of Phys Conf Ser 476 (2013) 012068*

IOP Conf. Series: Journal of Physics: Conf. Series **922** (2017) 012008 doi:10.1088/1742-6596/922/1/012008

Numerical model of a passive microsystem detecting and saving independent acceleration shocks

Karin Wedrich, Hannes Mehner, Martin Hoffmann

Technische Universität Ilmenau, Ilmenau, Germany.

karin.wedrich@tu-ilmenau.de

Abstract. This paper describes a numerical model for a passive microsystem to detect and save acceleration shocks. For the first time a system is presented, where more than one or two acceleration events can be detected and saved and where the latching position is not influenced by the seismic mass position before the acceleration event. Therefor the latching positon indicates the maximal external acceleration amplitude acted on the microsystem. A numerical model is used for prediction of the dynamic system behavior. Utilizing the model a parameter study was performed to determine optimal parameters for clear correlation between ratcheting position and its acceleration threshold. A mass of the ratcheting parts in the same range as the seismic mass enable a clear result in a range of up to 500 g and five latching positions.

1. Introduction

Shocks can lead to damages within e.g. the logistics chain. Also for supervision of product reliability and preventive maintenance, a detection of off-limit conditions is of interest [1]. To measure the maximum acceleration in this case, a passive microsensor is adequate. A passive sensor measures an environmental event without requiring electrical energy at that time. A non-electrical status change is utilized to store this event for an arbitrary long time. If the stored value is required the status can be read out e.g. by RFID.

The majority of passive acceleration sensors uses a spring-guided mass to detect the acceleration events [2-10]. A ratcheting mechanism stores the acceleration threshold for an arbitrary long time [2-5, 8-10]. The known systems contain one or two stable positions of the seismic mass [2,4-5,7-9] for detecting and storing one or two thresholds. In [10], a system is presented which features 20 ratcheting positions. The tooth shape secures the seismic mass position so that an acceleration event directed in the opposite direction of the sensitive direction of the microsystem leads not to a change of the seismic mass position. However, the acceleration thresholds for the next ratcheting position depends on the acceleration event itself as well as on the ratcheting position before the acceleration event. For most applications, it is important that the stored maximum value is independent from previous, smaller shock events.

The goal of this paper is a mathematical model for a system where the ratcheting position is independent from the ratcheting position and thus previous shock events before. So, every latching position can be correlated with a discrete acceleration amplitude range. The concept is derived to the idea in [10]. The parameters are optimized with respect to the system parameter *clearness,* which mean, that the seismic mass position does not influence the seismic mass position of the next acceleration event. The clearness of the latching position of the seismic mass in a passive microsystem is described by the first time.

IOP Conf. Series: Journal of Physics: Conf. Series **922** (2017) 012008 doi:10.1088/1742-6596/922/1/012008

2. Description of the mathematical model

The schematic system design is shown in figure 1. A seismic mass is coupled to the system frame by four guiding springs. They allow a movement of the seismic mass in the sensitive direction. Two stationary ratcheting parts are spring-guided too. The ratcheting mechanism is realized by numerous teeth between the three parts. The movement direction of the ratcheting parts is orthogonal to the seismic mass movement.

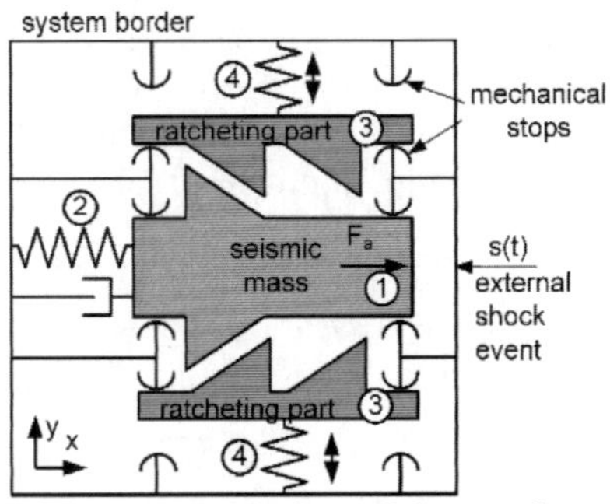

Figure 1. Basic system design of the passive shock sensor

In case of a sufficiently large acceleration amplitude, the seismic mass collides with the two corresponding ratcheting parts that are deflected due to this. If the seismic mass passes the first tooth, the measured acceleration was bigger than the first threshold. The tooth shape inhibits the backwards directed movement of the seismic mass. The number of passed teeth depends on the acceleration amplitude, only. The ratcheting parts are symmetrically arranged to avoid a seismic mass rotation around the z-axis during collision.

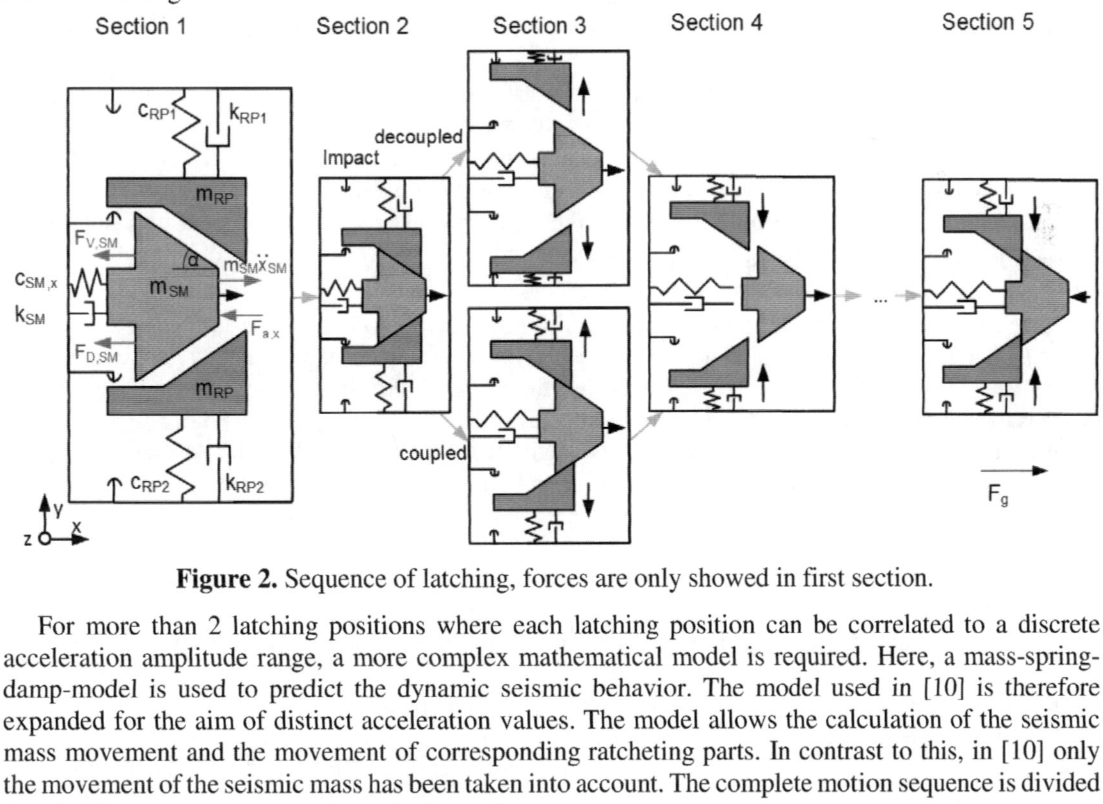

Figure 2. Sequence of latching, forces are only showed in first section.

For more than 2 latching positions where each latching position can be correlated to a discrete acceleration amplitude range, a more complex mathematical model is required. Here, a mass-spring-damp-model is used to predict the dynamic seismic behavior. The model used in [10] is therefore expanded for the aim of distinct acceleration values. The model allows the calculation of the seismic mass movement and the movement of corresponding ratcheting parts. In contrast to this, in [10] only the movement of the seismic mass has been taken into account. The complete motion sequence is divided into 5 different steps that are shown in figure 2.

In section 1, the seismic mass movement is described by the following equation in positive x-direction:

$$m_{SM}\ddot{x}_{SM} = \underbrace{m_{SM}a(t)}_{|F_{a,x}|} - \underbrace{c_{SM,x}x_{SM}}_{|F_{V,SM}|} - \underbrace{k_{SM}\dot{x}_{SM}}_{|F_{D,XM}|} \qquad (1)$$

F_{ax} is the force from the external acceleration event acting on the seismic mass in sensitive direction. $F_{V,SM}$ is the spring force described by the spring rate of the guiding springs of the seismic mass c_{SMx}, $F_{D,SM}$ is the damping force with the damping factor k_{SM}. The accelerations course is assumed to be ideally half sinusoidal-like in [10, 11] and as used in manufactures' instructions [12, 13].

In the second section, the seismic mass collides with the stationary ratcheting parts which can be described with the conservations of momentum and energy. Hence, the velocity of the seismic mass after collision is described by:

$$\dot{x}_{SM\,1,2} = \frac{\frac{m_{SM}^2}{2m_{PR}}\dot{x}_{SM}\left(1+\frac{1}{tan^2\alpha}\right) \pm \sqrt{\left(\frac{m_{SM}^4}{2m_{RP}^2tan^2\alpha}+m_{SM}^2\right)\dot{x}_{SM}^2+\left(2m_{SM}m_{RP}+2m_{SM}^2+\frac{m_{SM}^2}{tan^2\alpha}\right)\dot{y}_{RP}^2}}{2\left(\frac{m_{SM}}{2}+\frac{m_{SM}^2}{4m_{RP}}\left(1+\frac{1}{\tan^2\alpha}\right)\right)} \qquad (2)$$

In comparison to [10], the velocity of the ratcheting parts is calculated, too. However, for the first collision the velocity of the corresponding ratcheting parts is zero. The spring rate of the ratcheting parts in x-direction is estimated to be about 300 N/m, hence it can be assumed, that the ratchetings part do not move in x-direction. External acceleration on the ratcheting parts and the seismic mass in y- or z-directions are not noticed in this model. For applications like logistic monitoring or product reliability is just one acceleration direction dominant.

Dependent on the seismic mass velocity at the moment of collision, the seismic mass and the ratcheting parts move either coupled or decoupled in section 3. In case of a coupled movement, Coulomb friction occurs between the tooth flanks and has to be added to the differential equation of the seismic mass (1):

$$m_{SM}\ddot{x}_{SM} = m_{SM}a(t) - c_{SM,x}x_{SM} - k_{SM}\dot{x}_{SM} - 2\tan\alpha\frac{\sin\alpha + \mu\cos\alpha}{\cos\alpha - \mu\sin\alpha}(m_{RP}\ddot{x}_{SM} + k_{RP}\dot{x}_{SM} + c_{RP}x_{SM}) \qquad (3)$$

If the seismic mass and the corresponding ratcheting parts have different velocities after the collision, they are moving decoupled (1) in section 3. Consequently, the movement of the ratcheting parts has to be considered in the model in order to predict the moment and position of the next collision. Their movement can be described using a differential equation:

$$m_{RP}\ddot{y}_{RP} = m_{RP}a(t) - c_{RP}y_{RP} - k_{RP}\dot{y}_{RP} \qquad (4)$$

After a travel of about 5 µm in y-direction (additional to the tooth height) the ratcheting parts reach a mechanical stop. The stop is assumed to have an infinite mass. Hence, the velocity of the ratcheting parts becomes zero and they are pulled back by the spring force afterwards. The movement range towards the seismic mass is restricted by these stops as well.

In section 4 the seismic mass movement is calculated with the speed of the seismic mass after collision as initial conditions. If the ratcheting parts move decoupled from the seismic mass, the movement of the seismic mass is described by the same equation (1) from section 1. For the next collision there are two interesting facts to calculate: The velocity and movement direction of the seismic mass as well as the position of the next collision. This position is calculated considering the movement of the seismic mass and the ratcheting parts.

If the acceleration force is smaller than the retracting force caused by the seismic mass guiding springs, the seismic mass is slowed down, pulled backwards and stopped on the next foregoing tooth flank which represents the end of the ratcheting sequence (section 5).

The dynamic performance of the system is solved numerically by using the event function of the ODE45 solver in Matlab. Every section of the movement can be detected and the latching can be considered within the model.

IOP Conf. Series: Journal of Physics: Conf. Series **922** (2017) 012008 doi:10.1088/1742-6596/922/1/012008

3. System design

Based on this mathematic model, a system is designed that features 5 latching positions with a unique performance of *clearness*. That means every latching position can be dedicated to a specific acceleration range and does not depend on the seismic mass position prior to the next higher acceleration event. To achieve this behavior, the ratcheting parts have to be deflected suchlike that the ratcheting parts pass exactly one tooth.

A parameter study was executed to determine the influence of key system parameters on the ratcheting behavior. Key parameters are:

- the spring rate of the guiding springs for the seismic mass c_{SMx},
- the spring rate of the guiding springs for the ratcheting parts c_{RP},
- the mass of the ratcheting parts m_{RP},
- the ratcheting period b_Z,
- the ratio of the tooth length to the tooth height

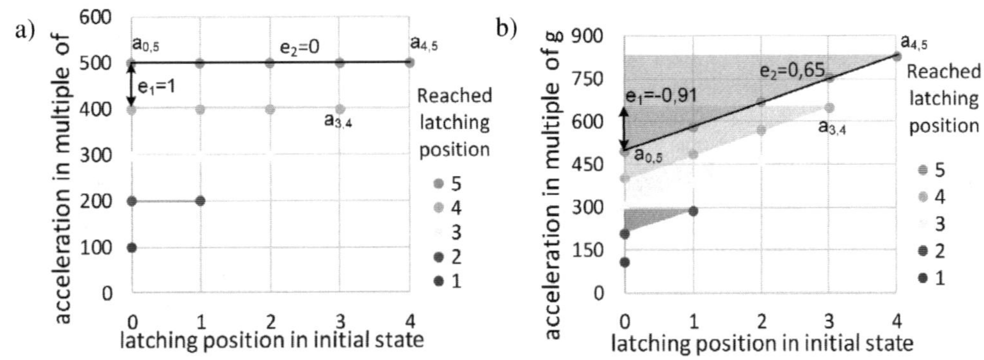

Figure 3. a) ideal clear system behavior; b) unclear system behavior

The seismic mass and the spring rate of its guiding springs show a similar influence on the system behavior. Due to the easy adjustability of the spring rate, the seismic mass is set as constant (0.3 mg). Parameters are successively varied in a reasonable technology range whereby the other parameters are constant. Subsequently the system performance is analyzed with 5 threshold-accelerations. As shown in fig. 3, clear system behavior comprises an equal distance between these 5 acceleration thresholds. An unclear system behavior is shown in figure 3 b). To calculate the clearness factor, two criteria are defined. The first criterion of distinctness is defined:

$$\text{clearness}_{distinctness} = e_1 = \frac{\frac{a_{0,5} - a_{3,4}}{a_{3,4}}}{\frac{a_5}{a_4} - 1} \tag{5}$$

a_5 is the desired acceleration amplitude for latching position 5. $a_{3,4}$ means the minimal necessary acceleration to reach the latching position 4 from the latching position in initial state 3. This is the main factor. The optimal value would be 1 as shown in figure 3 a). In case of an unclear behavior the value becomes negative or larger than 2.

The second criterion using the gradient of the threshold-acceleration on latching position 5 is calculated with the equation:

$$\text{clearness}_{rise} = e_2 = \frac{a_{4,5} - a_{0,5}}{a_{0,5}} \tag{6}$$

The optimum for this criterion is 0. Both criteria are influencing each other as shown figure 3 b). The influences of the investigated parameters are shown in figure 4 a). The influence was determined by taking the standard deviation from varied parameters in a reasonable range of technology.

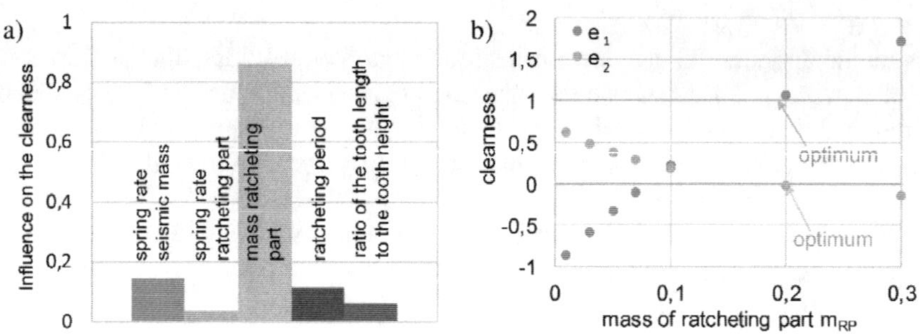

Figure 4. a) Influence of the parameter on the clearness; b) Influence of the mass of the corresponding ratcheting parts with $m_{SM} = 0.3$ mg, $c_{SMx} = 2$ N/m, $c_{RP} = 30$ N/m, $b_Z = 200$ µm, $\alpha = 30$ °, $a_Z = 35$ µm

An increase of the parameter c_{SMx} causes an increase of all acceleration thresholds but has just a little influence on the clearness. The spring rate of the guidance of the ratcheting part c_{PR} has nearly no influence on the system performance due to the uncoupled movement for the majority of acceleration events. The mass of the ratcheting parts has great influence on the system behavior (shown in figure 4 b)). The greater the mass of the ratcheting parts, the more energy of the seismic mass is transferred to the ratcheting parts. In figure 4 b) the optimum of the ratcheting parts is shown. The influence of the latching period is similar to the influence of the parameter c_{SMx}. The parameter latching period, which contains the travel of the seismic mass and the parameter c_{SMx} are linear components in the spring force. So, the mass of the ratcheting part has the biggest influence of the parameter *clearness*.

For verifying the presented model demonstrator systems will be fabricated, which can detect accelerations until 500 g (shown in figure 5). The system parameters are summarized in table 1. The clearness factor distinctness is 1.1, which is close to the optimum 1 and the gradient of threshold-acceleration is -0.2 %, close to the optimum 0. The gradient of the other threshold-acceleration is about 10 %. The first threshold-acceleration is by 73 g.

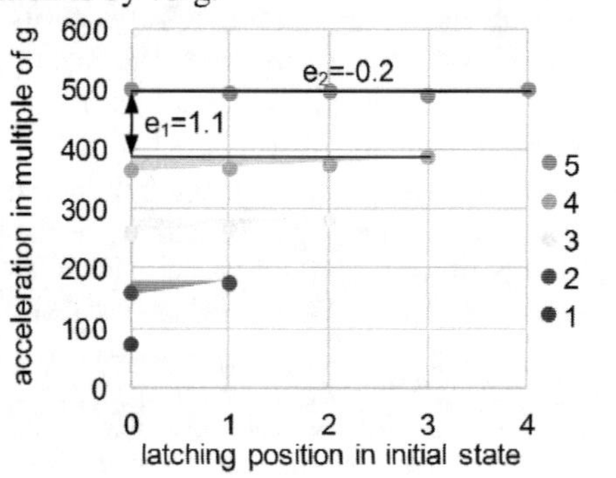

Figure 5. Clearness of shock sensor with the parameters: $m_{SM} = 0.75$ mg, $c_{SMx} = 7.37$ N/m, $m_{RP} = 0.6$ mg, $c_{RP} = 12.5$ N/m, $b_Z = 100$ µm, $a_Z = 35$ µm

Table 1. Key demonstrator parameters

Parameter	**Symbol**	**Unit**	**Assumed Value**
Movement range (ratcheting period)	x_{max} (b_z)	µm	500 (100)
Number of ratcheting positions	n		5
Spring rate of the guiding springs for the seismic mass	c_{SMx}; c_{SMz}	N/m	7.4; 30 -7.
Seismic mass	m_{SM}	mg	0.75
Spring rate of the guiding springs for the ratcheting part	c_{RP}	N/m	12.5
Mass ratcheting part	m_{RP}	mg	0.6
Tooth shape	a_Z; h_Z; α_Z	µm or °	52.5; 30; 30

By having an optimum by other maximum accelerations, the ratio of the seismic mass and the mass of the ratcheting parts change. While by 500 g maximum acceleration the ratio is about 1:0.8, the ratio by 5000 g is about 1:0.2.

4. Conclusions and further work

A numerical model for a passive acceleration threshold detector and the influences of the system parameters on its behaviour were shown. The parameters are optimized to have a distinct measurement. That means, that every latching position a range of acceleration can be collated. Of course a lower seismic mass and a higher stiffness of the guiding spring of the seismic mass leads to a system which senses higher accelerations. The spring rate of the guiding springs of the ratcheting parts have very low effect on the behaviour of the seismic mass. The strongest influence of the parameter clearness has the mass of the ratcheting parts, because the energy of the seismic mass is transferred to the ratcheting parts by impact. A mass of the ratcheting parts in the same range as the seismic mass enable a clear result in a range of up to 500 g and five latching positions.

For verifying the presented model the next step is the fabrication of a demonstrator system with the shown parameters from table 1. In an adequate measurement setup the behavior of the seismic mass will be reviewed.

References

[1] M. R. Whitley, M. Kranz, R. Kesmodel, S. Burgett, Latching Shock Sensors for Health Monitoring and Quality Control, MEMS/MOEM S components and their applications II, Proc SPIE 5717 January 24-25, 2005, pp. 185-194.

[2] Ciarlo, D.R., A latching accelerometer fabricated by the anisotropic etching of (110) oriented silicon wafers. Micromech Microeng, 1192. 2.

[3] Xi-Qing Sun, S.Z.a.W.N.C., A Surface Micromachined Latching Accelerometer, in International Conference on Solid-State Sensors and Actuators. 1997: Chicago.

[4] Sung-Yueh Wu, C.-Y.H., Wensyang Hsu, A wirelessly readable and resettable shock recorder through the integration of LC circuits and MEMS devices. Smart Materials and Structures 2014. 23.

[5] B J Hansen, C.J.C., B D Jensen, A R Hawkins, S M Schultz, Plastic latching accelerometer based on bistable compliant mechanisms. Smart Materials and Structures, 2007. 16.

[6] Benjamin Todd, M.P., Stephen M. Schultz, Aaron R. Hawkins, Brian D. Jensen Low-Cost RFID Theshold Shock Sensor. IEEE Sesors Journal, 2009. 9(4).

[7] A. Frangi, B.D.M., F. Confalonierei, S. Zerbini, Threshold Shock Sensor Based in a Bistable Mechanism: Design, Modeling, and Measurements. Microelectromechanical Sys., 2015. 24(6).

[8] L. J. Currano, S.B., W. Churaman, M. Peckerar, J. Wienke, S. Kim, M. Yu, B. Balachandran, Latching ultra-low power MEMS shock sensor for acceleration monitoring. Sensors and Actuators A: Physical, 2008.

[9] Z Y Guo, Q.C.Z., L T Lin, HT Ding, XS Liu, J Cui, Z C Yang, Huikai Xie and G Z Yan, An acceleration switch with a robust latching mechanism and cylindrical contacts. Journal of Micromechanics and Microengineering, 2010. 20.

[10] H. Mehner, S. Schwebke, S. Leopold, M. Hoffmann, Passive microsensor for binary counting of numerous threshold events, Proc. SPIE 9517, Smart Sensors, Actuators, and MEMS VII; and Cyber Physical Systems, (2015).

[11] Ch. Lalanne Mechanical Shock- Mechanical Vibration and Shock Volume 2 1999 ISBN 1 90399604 X

[12] Hewlett-Packard Development Company, L.P. (Hrsg.): HP Compaq 6735b Notebook PC and HP Compaq 6730b Notebook PC Maintenance and Service Guide. L.P., Juli 2008

[13] Dell, Inc. (Hrsg.): Dell Studio 1555/15 setup guide. Dell, Inc., August 2009

IOP Conf. Series: Journal of Physics: Conf. Series **922** (2017) 012009 doi:10.1088/1742-6596/922/1/012009

Non-resonant energy harvester with elastic constraints for low rotating frequencies

Sebastián P Machado[1], **Mariano Febbo**[2], **Claudio D Gatti**[1] **and José M Ramirez**[1]

[1] Grupo de Investigación en Multifísica aplicada, Universidad Tecnológica Nacional FRBB and CONICET, Bahía Blanca, Argentina.

[2] Instituto de Física del Sur, Departamento de Física. Universidad Nacional del Sur, CONICET, Bahía Blanca, Argentina.

smachado@frbb.utn.edu.ar

Abstract. This paper presents a non-resonant piezoelectric energy harvester (PEH) which is designed to capture energy from low frequency rotational vibration. The proposed device works out of the plane of rotation where the motion of a mass-spring system is transferred to a piezoelectric layer with the intention to generate energy to power wireless structural monitoring systems or sensors. The mechanical structure is formed by two beams with rigid and elastic boundary conditions at the clamped end. On the free boundaries, heavy masses connected by a spring are placed in order to increase voltage generation and diminish the natural frequency. A mathematical framework and the equations governing the energy-harvesting system are presented. Numerical simulations and experimental verifications are performed for different rotation speeds ranging from 0.7 to 2.5 Hz. An output power of 125 μW is obtained for maximum rotating frequency demonstrating that the proposed design can collect enough energy for the suggested application.

1. Introduction

This paper presents an innovative design of a PEH system which represents an alternative to the most common device consisting in cantilevered beams carrying tip masses to generate energy on a rotating environment. Most of previous works [1-3] considered piezoelectric beams fixed to a rotating hub, where the effect of the centrifugal force is used to tune the resonance frequency with the rotating driving frequency. When the hub rotates at constant angular speed, the output power and voltage of the harvester occur at the resonance frequency. In ref. [1], the authors only presented the utilization of the FE method to simulate a passive self-tuning piezoelectric composite beam for harvesting rotational energy. They obtained a relatively good power generation but their model failed to predict a precise peak voltage generation. A multimode mathematical model for the coupled electromechanical system is derived in [2], but experimental results are not presented. A simple analytical model is presented in [3] to explain the design parameter selection. The prototype reached a passive self-tuning for driving frequencies ranging from 6.2 to 16.2 Hz.

According to these proposals [1-3], it is required long length scales or large masses to match the resonance frequency to the excitation frequency for low rotation speeds. In order to overcome this difficulty, we propose a two-beam structure connected by a mass-spring system supported by elastic constrains. The device is intended to deliver energy for wireless structural monitoring systems or sensors, to monitor wind turbines of 30 KW working at 50~150 rpm. These characteristics impose a definite bandwidth and size to the EH device to be constructed. For this reason, the physical dimensions and geometry are determining factors in the design.

2. Device Concept

A schematic illustration of the device is shown in figure 1, where it can be noticed that the harvester is mounted rigidly on a hub. The gravity component in the reference system fixed to the hub, can be used in rotary motion to cause a beam to vibrate harmonically in the transverse direction, similarly to a base excitation. The advantage of the proposed shape over that of a simple cantilever beam can be explained by means of natural frequency analysis. The structural system is formed by two beams, which have very different stiffnesses. The harvester design is based on reducing the natural frequency by means of elastic restraints adopted for beam #2 (figure 1). The piezoelectric sheet (QP16N, Mide Corporation) is mounted on the stiffest beam #1. Both beams are connected by a spring element (spring #1) placed between two heavy masses. For these reasons, the configuration of the system requires a complex analysis of the natural frequencies and mode shapes, which will be presented in the next sections.

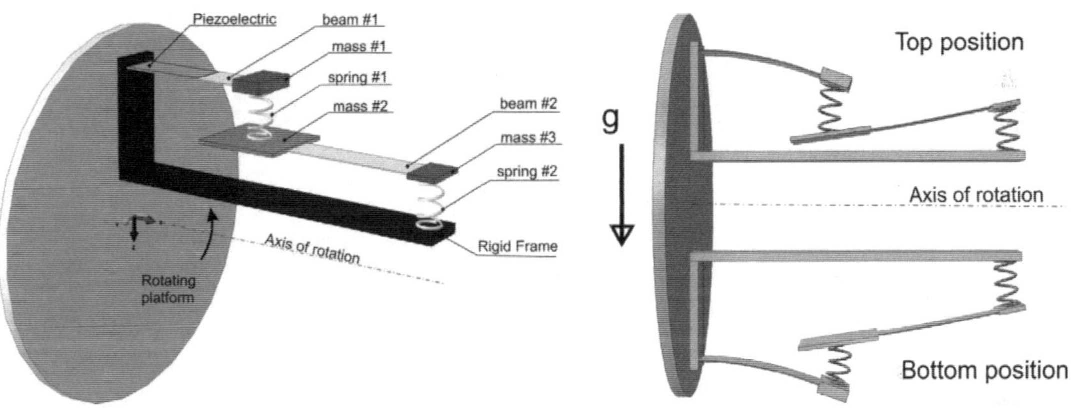

Figure 1. Schematic views of the proposed PEH.

3. Analytical model

The mathematical modeling of the proposed mechanism is simplified assuming that Euler-Bernoulli beam equations are valid for both beams. A Lagrangian approach is then used to develop the governing equations. The position of an infinitesimal segment on the beam in the rotating coordinate system is represented by R:

$$R = (r + w)e_z + u\ e_x, \tag{1}$$

where w and u are the transverse and axial elastic displacements respectively, r is the distance from the axis of rotation to the beams, while e_x and e_z are unit vectors in the x and z directions respectively. In this case the bending motions will not induce significant axial motions for that reason they are neglected. The kinetic energy of the system is

$$T = \frac{1}{2}\rho A \int_0^l \left[\dot{w}^2 + \dot{\theta}^2 (w^2 + r^2 + 2r\ w) \right] dx + \frac{1}{2} m_T \dot{R}t_l \cdot \dot{R}t_l + \frac{1}{2} I_T \dot{w}_l'^2, \tag{2}$$

where l is the structural system length, m_T represent the masses (#1, #2 or #3), I_T is the moment of inertia of the masses and the velocity of the heavy masses at free end is

$$\dot{R}t = (\dot{w} + d\ \dot{w}_l')e_x + (r + w_l + d\ w_l')\dot{\theta}\ e_z \tag{3}$$

IOP Conf. Series: Journal of Physics: Conf. Series **922** (2017) 012009 doi:10.1088/1742-6596/922/1/012009

where d is the length from the tip beam to the center of the heavy mass (see figure 2). The potential energy of the system can be expressed as

$$U = \frac{1}{2} EI \int_0^l w''^2 \mathrm{d}x + g\rho A \int_0^l (r+w)\, sen\theta \,\mathrm{d}x + m_T g \left[(r + w_l + d\, w_l')\, sen\theta \right] - \frac{1}{2} J_p v \int_0^l w'' \mathrm{d}x \quad (4)$$

where J_p is the electromechanical coupling coefficient, v is the voltage drop in the piezoelectric, EI is the flexural rigidity which depends on the section of the structural system and g is the gravity. The internal electrical energy in the piezoceramic layer is

$$W_{ie} = \frac{1}{2} J_p v \int_0^l \frac{\partial^2 w}{\partial x^2} \mathrm{d}x + \frac{1}{2} C_p v^2 , \quad (5)$$

being C_p the internal capacitance of the piezoceramic. Applying the Lagrangian formulation for the mechanical and electrical degrees of freedom it is possible to define the governing equations of motion in the following way:

$$\ddot{q} + c\dot{q} + (\mathrm{K} - \Omega^2) q - \Omega^2 \chi + \vartheta\, v = \Gamma g \sin\Omega , \quad (6)$$

$$C_p \dot{v} + \frac{v}{R} + \vartheta\, \dot{q} = 0 . \quad (7)$$

In order to consider the geometric change introduced by the piezoelectric layer and the spring connection, a discontinuous system is proposed to define an orthogonal basis function (see figure 2). With the aim to show the complex analysis of the natural frequencies and mode shapes, the modal amplitude constant is evaluated from a mass normalization according to:

$$\begin{aligned} &\rho A_1 \int_0^{L_1} \phi_{j,1}(x)\phi_{n,1}(x)\,\mathrm{d}x + \rho A_2 \int_{L_1}^{L_1+L_2} \phi_{j,2}(x)\phi_{n,2}(x)\,\mathrm{d}x + \rho A_3 \int_{L_1+L_2}^{L_1+L_2+L_3} \phi_{j,3}(x)\phi_{n,3}(x)\,\mathrm{d}x + \\ &\phi_{j,2}(L_1+L_2) M_1\, \phi_{n,2}(L_1+L_2) + \phi_{j,3}(L_1+L_2) M_2\, \phi_{n,3}(L_1+L_2) + 2M_1 d_1 \phi_{j,2}(L_1+L_2)\, \phi'_{n,2}(L_1+L_2) + \\ &2M_2 d_2 \phi_{j,3}(L_1+L_2)\, \phi'_{n,3}(L_1+L_2) + (I_{T1} + M_1 d_1^2)\phi'_{j,2}(L_1+L_2)\phi'_{n,2}(L_1+L_2) + \\ &(I_{T2} + M_2 d_2^2)\phi'_{j,3}(L_1+L_2)\phi'_{n,3}(L_1+L_2) = \begin{cases} 0 \text{ for } j \neq n. \\ 1 \text{ for } j = n. \end{cases} \end{aligned} \quad (8)$$

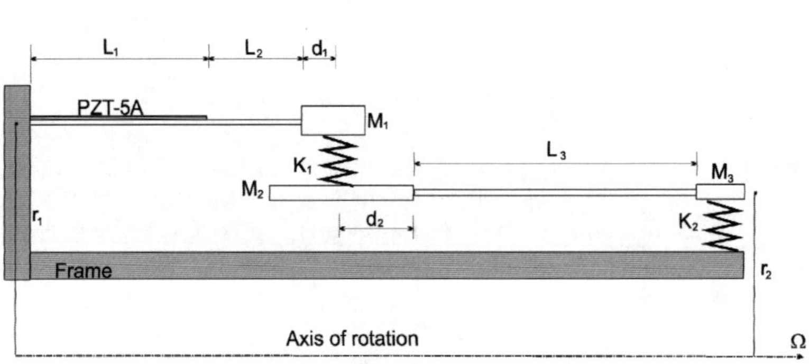

Figure 2. Schematic showing the parameters of the PEH.

For a complete derivation of these expressions, the more involved reader can follow ref [5].

4. Experiments

To investigate the validity of theoretical model, a prototype device has been fabricated and tested. An electric motor with a variable speed controller provides the rotational motion of the system which was rigidly mounted to the hub as observed in figure 3. Instead of using a more common slip-ring system which is a major shortcoming to acquire the voltage signal in a rotating scenario, an Arduino board with a Bluetooth connection was used instead, that connects the collected data to a pc at a rate of 100 samples/sec. The design parameters of the harvester are summarized in table 1 and the other parameters are: $K_1 = K_2 = 4300$ N/m, $M_1 = 29$ gr, $M_2 = 42$ gr, $M_3 = 10$ gr, $d_1 = 8$ mm, $d_2 = 22.5$ mm, r_1

=90 mm and r_2 =65 mm. A load resistor of 56 KΩ was employed in order to capture the voltage generation, due to a limitation of the data acquisition system (Arduino setting= 5 Vpp max).

Table 1. Design parameters of the piezoelectric harvester.

	Length (mm)	Width (mm)	Thickness (mm)	Density (kg/m^3)	Young`s modulus (GPa)	Coupling d_{31}(pm/V)	Capacitance (nF)	Permittivity
Beam #1	73	22.5	0.5	7850	210	-	-	-
Beam #2	70	18	0.5	2700	70	-	-	-
QP16N	45.9	20.57	0.25	7800	67	-190	90.78	1500 ε_0

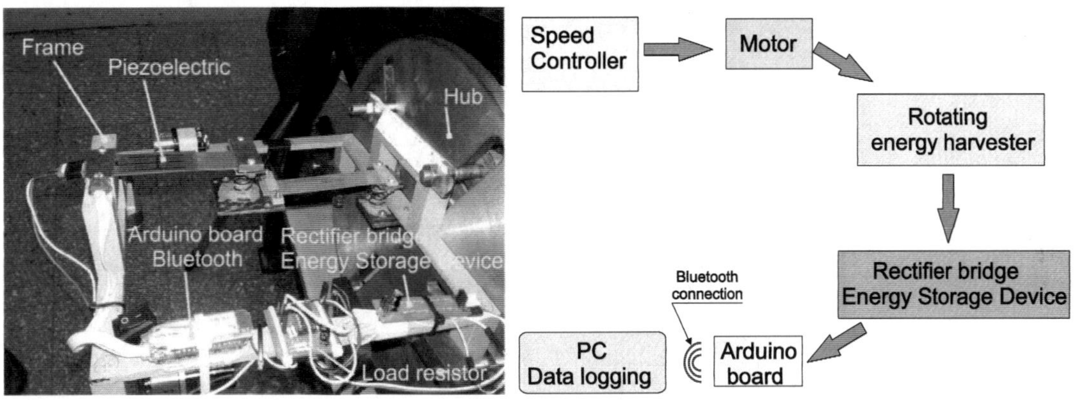

Figure 3. Prototype and experimental setup of the PEH.

5. Results

5.1. Frequency analysis

As it was pointed out before, we aim to reduce the natural frequency of the PEH by means of a proper design of beam #2 (secondary beam). We begin by examining the natural frequencies for beams #1 and #2. In a cantilever condition without spring connection (K_1=0), the frequencies are $f_{\#1}$= 22.8 Hz and $f_{\#2}$= 6.7 Hz. When coupled to spring/mass system, the combined effect of the two beams provides a new frequency value of f = 13.8 Hz. In this way, the frequency of the PEH is reduced considerably by means of the two elastic conditions K_1 and K_2. As the system rotates, there is an additional reduction in frequency due to the centrifugal force, but this reduction is slight compared with the first, being from 13.8 →13.4 Hz for a speed of Ω = 150 rpm.

5.2. Rotational harvester: Test and modelling

The dynamic behavior of the energy harvesting system under rotational motion is shown in this section. In figure 4 the peak voltage generation obtained by analytical and experimental means are presented for rotation speeds between 50~150 rpm. As it can be found, experimental and theoretical results match very well for the whole set of analysed frequencies. The output voltage ranges from a minimum value 0.54 Volts to a maximum of 1.93 Volts with a linear increment for growing frequencies.

With the aim to increase the generated voltage, we add an extra mass of 31 gr to M_2 (see figure 1). This value was selected in order to ensure a geometrically linear assumption, as will be demonstrated later. We expect this extra mass has a twofold effect, on the one hand to increase the strain on the beam #1 and therefore the bending moment; on the other hand, to diminish the natural frequency of the PEH. Figure 5 shows the results.

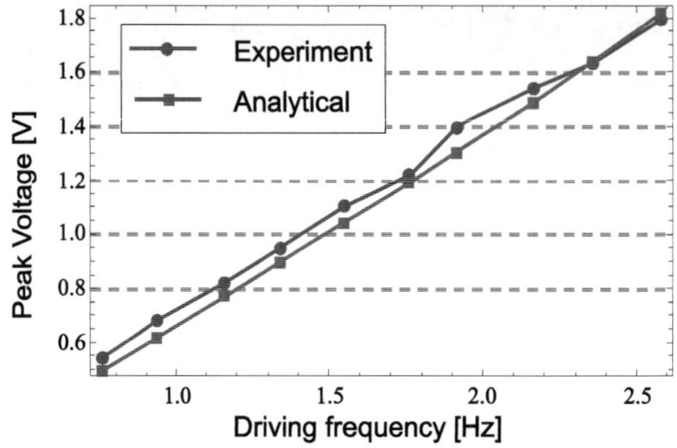

Figure 4. Peak output voltage versus rotating frequency.

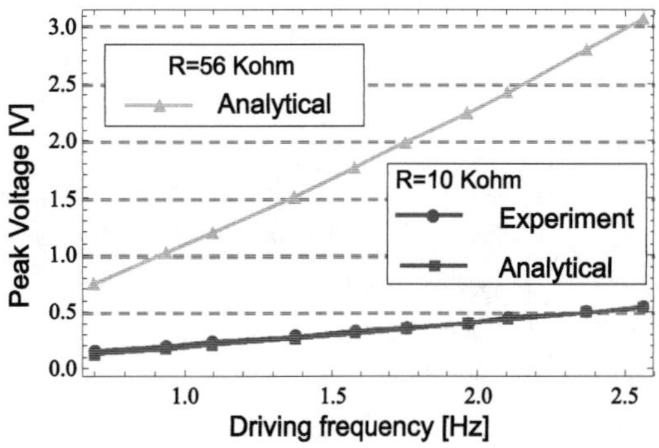

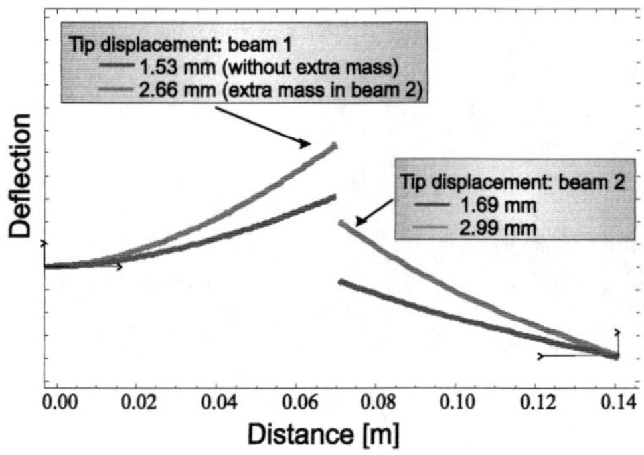

Figure 5. The influence of extra mass on the voltage. **Figure 6.** Comparison of tip displacement.

6. Discussions

As it can be observed, the output voltage was increased by 66% and the resonance frequency of the harvester was measured in 10.86 Hz (compared to 13.8 Hz). Clearly, this mass addition confirmed both benefits on increasing voltage generation. Is it important to point out that the results for R=56 KΩ are only analytical predictions due to the limitation of the data acquisition system. Instead, we diminish the resistive load to R=10 KΩ to present a proper comparison between experimental and theoretical predictions.

Figure 6 shows the maximum deflection of the beams for the largest rotation speed (150 rpm). There, it is possible to observe that the ratio of tip displacement to length of the beams is below 0.1, demonstrating the validity of the linear assumption. It is also interesting to note that the large displacements observed in the case with extra mass, confirm that the strain induced in the beam #1, which supports the piezoelectric layer, is larger compared with the case of no extra mass.

In order to accumulate the harnessed power, a simple harvesting circuit, which consists of a rectifier bridge and a 47 μF electrolytic capacitor acting as a storage device, was used. We select germanium diodes instead of silicon ones in the rectifier bridge to minimize the voltage drop. Therefore, the capacitor accumulates the rectified DC generated by the REH as the system rotates. To calculate the average harvested power during a certain time period, we follow ref [4], with V_{rect1} = 5V and V_{rect0} = 4V and the time interval Δt computed from the collected data. The results are presented in figure 7 for the cases with and without an extra mass in beam #2. From these results, it can be observed a minimum harvested power of 34.85 μW and maximum of 123 μW for a rotating frequency of 2.56 Hz (153.6 rpm) and a minimum of 34 μW for 0.69 Hz (41.4 rpm) in the case of the addition of mass in beam #2. Taking into account that a minimum power requirement of 20 μW is needed to feed a wireless transmitter circuit [4], our system can be used as a power source for autonomous structural monitoring systems or sensors. It is important to point that having used a more involved harvesting circuit, the output power could have been increased [4]. This is left for future works.

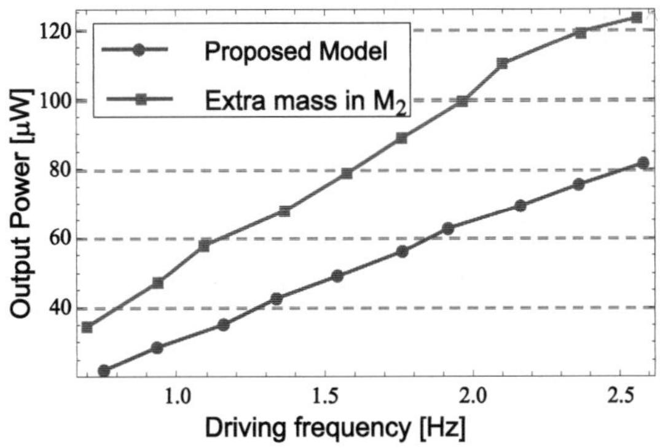

Figure 7. Output power versus rotating frequency.

7. Conclusion

This paper presents the design of a piezoelectric energy harvester to capture energy from low rotating frequencies. The objective was to increase the output voltage by studying the interaction between its constitutive parts which are comprised by two beams and a spring-mass system supported by elastic boundary conditions. The selection of these boundary conditions, in comparison with cantilever beam models, lower the stiffness of the entire harvester without enlarging the length of the beams. By the addition of some extra mass in a particular place of the device, it is also possible to diminish its natural frequency and to increase the strain induced in the piezoelectric sheet, rising in this way the generated power. A thorough mathematical (electromechanical) model for the harvester was developed and validated with experimental tests. Numerical simulations and experimental verifications are performed with very good agreement for different rotation speeds ranging from 0.7 to 2.5 Hz. With a generation between 34.85-123 μW, the results demonstrate that it is possible to use this system to power up wireless sensors for the projected application.

Acknowledgements
The authors wish to thank CONICET (PIP Nº:11220120100346), UNS, UTN FRBB and ANPCyT under grant PICT 2013- Nº 2065 for their financial support.

References

[1] J. C. Hsu, C. T. Tseng and Y. S Chen, "Analysis and experiment of self-frequency tuning piezoelectric energy harvesters for rotational motion," *Smart Mater. Struct.*, vol. 23, no. 075013, 2014.
[2] F. Khameneifar, M. Moallem and S. Arzanpour, "Modeling and Analysis of a Piezoelectric Energy Scavenger for Rotary Motion Applications, " *J. Vib. Acoust.*, vol. 133, no. 011005, 2011.
[3] L. Gu and C. Livermore, "Passive self-tuning energy harvester for extracting energy from rotational motion," Appl. Phys. Lett., vol. 97, no. 081904, 2010.
[4] M. Guan and W. H. Liao, "Design and analysis of a piezoelectric energy harvester for rotational motion system," *Energy Convers. Manage.*, vol. 11, pp. 239-244, 2016.
[5] S. P. Machado, M. Febbo, F. Rubio-Marcos, L. A. Ramajo and M. S. Castro, "Evaluation of the performance of a lead-free piezoelectric material for energy harvesting", *Smart Mater. Struct.*, vol. 24, no. 115011, 2015.

IOP Conf. Series: Journal of Physics: Conf. Series **922** (2017) 012010 doi:10.1088/1742-6596/922/1/012010

Flowing and pressurizing a solid-liquid two phase monodispersed fluid with high solid content in a transparent microfluidic high-pressure chip

M Andersson[1], G Mårtensson[2,3] and L. Klintberg[1]

[1]*Department of Engineering Sciences, Division of Microsystems Technology, Uppsala University, Uppsala, Sweden*
[2]*Mycronic AB, Täby, Sweden*
[3]*EMSL, MC2, Chalmers University of Technology, Göteborg, Sweden*

Abstract. Handling highly concentrated solid-liquid two-phase fluids in microfluidics is challenging. In this paper, we present the first studies of flowing solder paste with a high solid content in a transparent high-pressure tolerant glass chip, thereby increasing the understanding of how multiphase liquids with high density difference between the phases behave in small channels (840 µm in diameter). The system, including a custom made high-pressure, low resistance, interface, was continuously operated at pressures up to of 6 MPa and devices where shown to have pressure tolerance up to 17 MPa. During flow through the chip, the packing density of the solder balls displayed inhomogeneity over the channel where chains of solder balls in contact with each other were formed together with voids. These in-homogeneities persisted along the channel during flow. The flow rate of the paste through the chip oscillated between 63 to 350 µm/s when pumping at constant volume rate of 30 µl/min. When a pressure of 2 MPa was applied, the volume of the solder paste particle segment decreased 1.6%, and 0.1% was elastically recovered when the pressure was released. It is concluded that this transparent microfluidic high-pressure glass chip with the special developed interface is suitable for flow studies of solder paste with a high solid content.

1. Introduction

One of the remaining major challenges in microfluidics is the handling of solids [1]. Both in technological and biological systems, concentrated suspensions are of vast importance. Blood, samples from organic matter, inorganic slurries, some drugs and paints, are a few examples of two-phase systems containing solid particles in a liquid [2]. Diluted solid-liquid microfluidic systems have been studied extensively [3], where particle focusing [4], trapping [5] and aligning the particles [6] are ways to manipulate the fluids. However, very few studies have been made using high solid content fluids in microfluidics. Our aim with this paper is to show how pumping of a concentrated, high viscosity solder paste suspension can be done in a microsystem and to highlight some of the difficulties that are necessary to deal with.

Solder paste is used to connect surface mounted components to printed circuit boards. The fluid is very tacky since one purpose is to temporarily keep the component in place before the actual soldering. The paste is a mixture of micrometer-sized metallic solder balls suspended in liquid flux. The non-Newtonian pastes are formulated in order to be highly thixotropic, meaning that the viscosity is lowered when the paste is sheared [7, 8]. This is a property seen in many fluids of industrial value, including paint, allowing for the liquid to flow readily while keeping its shape after being dispensed. The viscosity can also be very temperature dependent. Viscoplastic materials, like solder paste, mud

and concentrated slurries are also Bingham plastics, *i.e.* the stress needs to be above a yield stress before the fluid starts to flow [2]. This is a behavior found in concentrated solid-liquid two-phase systems, when there are large interactions between the particles in the liquid, forming a weak continuous solid structure.
To dispense the paste to the solder pads on printed circuit boards, stencil printing, pneumatic dispensing or jet printing can be used [9]. Jetting, being both flexible and fast, has become a standard process in surface mount technology production [10]. The size of the solder particles in jettable pastes are, depending on grades, in the range 2 to 35 µm, and the solid loading is typically 84-86% by weight. During jetting through orifices that can be down to 100 µm in diameter, the paste is subjected to extreme stresses with pressures up to 15 MPa and shear rates up to 10^{-6} s^{-1}. Since solder paste in many aspects is a very difficult fluid to handle, many details of the behavior of the movements in the microchannels inside the jet head is still unknown, making it hard to model and optimize the jetting.
In this paper, we present a high pressure tolerant transparent microfluidic glass chip designed for studies of the rheological behavior and compressibility of solder paste in order to obtain a deeper understanding on how multiphase liquids with a high solid content and high density difference between the phases behave in a microfluidic chip. Since the flow resistances and viscosities are much higher than from what normally is seen in microfluidics, we utilize large cross-sections channels and interfaces for our study.

2. Experimental details

To enable the flow of solder paste, a microfluidic chip and special interface must be designed, allowing for low flow resistance. First, an initial flow study using a syringe pump (Harvard, PHD 2000) was conducted using centimeter long tubing with an inner diameter ranging from 75 to 810 µm to determine what minimal dimensions that are needed for the solder paste (experimental sample, Mycronic) to flow through. The volume fraction of solid particles in the solder paste was around 0.4 and the particles had an averaged diameter of 5 µm. Based on the results from this study, a microfluidic chip having a T-junction and meander was constructed, Fig. 1. The chip was used to study the flow velocity, as well as investigating the response to the paste upon pressurization.
Glass chips were fabricated by the methods described in [11]. In short, borosilicate glass wafers were first sputtered with an etch mask made from a 2 µm thick Mo layer. This mask was patterned by lithography to create 60 µm wide openings followed by etching of the Mo with a solution of H_3PO_4:HAc:HNO_3:H_2O at a ratio of 180:11:11:150, thereby defining the channels. Using concentrated HF, trenches in the glass are isotropically etched down to 420 µm. After bonding two structured wafers together, the semi-circular cross-sections of the trenches form an almost circular channel. Before bonding, the glass was activated using 10 wt% KOH at 65°C followed by 69 wt% HNO_3 at 80°C. Bonded wafers were thermally treated at 625°C.
A low flow resistance fluid interface was specifically designed to connect to the chip. The interface consisted of 5 cm of stainless steel tubing (OD 1/16", ID 810 µm, IDEX) that was joined closely to the chip by inserting 6 mm long PEEK capillaries (OD 1/32", 800 µm, ID 510 µm, IDEX) into both the tube and chip inlets. A two-component epoxy (Araldite Rapid, Huntsman Advanced Materials) was applied around the interfaces of the tube and chip, forming a sealed and rigid junction. The two outlets marked b and c, Fig. 1, were then connected to two high-pressure needle valves (15-11AF, HIP). The valves were oriented so that when they were in a closed position, the needle cuts the fluid path from the chip perpendicular, thus minimizing dead volume in the valve.
To investigate the durability of the chip, a fracture test was done by inserting a silica capillary (150 µm OD, 75 ID, IDEX) into a straight segment of a channel and plugging the outlets with epoxy. Ethanol with a flow rate of 200 µl/min was pumped (ISCO 100 DM, Teledyne) into the chip by the capillary, resulting in an average pressure increase of 2.1 MPa/minute, read by the internal pressure sensor of the high-pressure pump.
For the flow study, a syringe pump (Harvard, PHD 2000) was used, set to 30 µl/min. For the flow studies, the temperature of the chip was raised to 27-32°C to decrease the viscosity of the paste and allow for pumping. A low friction syringe (1005, Hamilton) was filled with the solder paste.

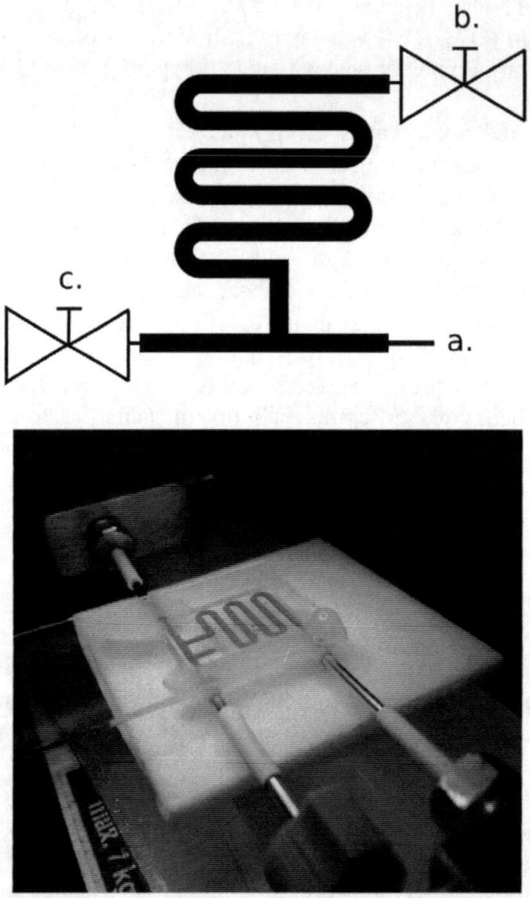

Figure 1. A sketch showing the channel geometry (top) and a mounted chip with fluid inlets and solder paste flowing in the channels (bottom). The inlet (a) is connected to the pumps, (b) and (c) are outlets connected to high-pressure needle valves. The high-pressure tolerant interfaces consisting of orange PEEK and stainless steel tubing are visible in the photo.

To investigate the effect of pressurizing the solder paste, a paste segment with a defined volume was first made. Paste was pumped in the chip from point a to b (Fig. 1) at 5 µl/min for 10 min, then the valve at point b was closed thereby cutting the paste segment off. By flowing hydraulic oil (HDZ 68, Texaco) from point a to point c, a defined segment of solder paste was formed having oil-paste interface at the T-junction. Using this procedure to fill the chip with paste, it was ensured that there were no air pockets present in the channel. While filling the chip with solder paste, the flow was visually monitored to ensure that no air segments were present in the fluid.

The volume, including tubing and channels in the chip up to the T-junction, was 5.6 µl. By closing the valve at point c and pressuring the chip using a high-pressure piston pump (ISCO 100 DM, Teledyne), the compression and the following the displacement of the paste segment could be studied. The temperature during the compression test was 21 ± 1 °C. The pressure was recorded using the internal pressure sensor of the high-pressure piston pump. Imaging and video recording were done with a stereoscope (Nikon SMZ800) with ring light coupled to a CCD camera (DFKMN33UXE174, Imaging Source). Image analysis of the videos was used to calculate the flow rates using the solder balls to track the flow velocity.

3. Results and Discussion

The durability test showed that the chip fractured at a pressure of 17.3 MPa and inspection of the cracked chip revealed that the crack propagated close to the bond plane inside the glass and in close proximity to the glued interface of the inlet. During the flow and compressibility tests, a maximum pressure of 6 MPa was applied.

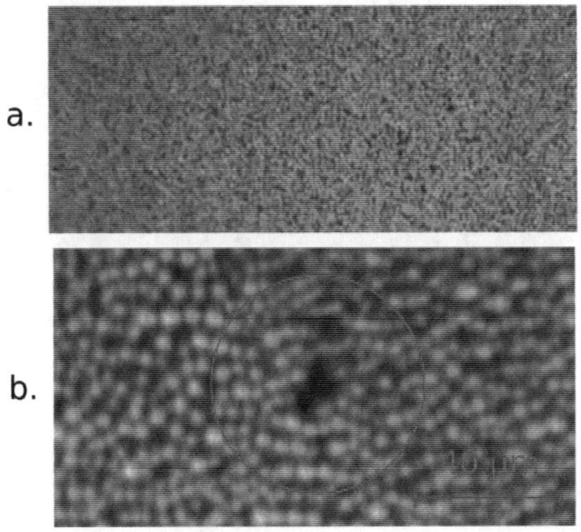

Figure 2. The solder paste flowing in the channel at 30 µl/min. In (a) an overview of the channel width is seen. (b) shows a close up, showing the solder balls. The solder balls are generally in an unordered glassy state, but local ordering in the form of chain structures can be seen. A void having no solder balls is marked with a red circle.

The initial trials lead to the conclusion that solder paste at room temperature could not be reliably pumped through tubes having an inner diameter smaller than 160 µm, and for the chip, the smallest diameter of the system was that of the interfacing tubing. This diameter was chosen as 510 µm, in order to not restrict the flow too much.
For the flow study, the resulting flow speed was not constant, but showed pulsing with faster flow periods, having a frequency of 0.07 +/- 0.007 Hz, lasting on average 15 s for each pulse. The flow velocities for fast and slow periods were 350 and 63 µm/s, respectively.
As Fig. 2 shows, the packaging of the solder balls is not perfectly homogeneous and voids, typically 5 to 30 µm in diameter, where no balls can be found are present. These voids continue to travel with the balls along the channels for long distances and can be observed for several seconds over the recorded image area. It was also observed that these voids can disappear, possibly moving into the bulk of the fluid. No gas-liquid interfaces can be seen in the voids, suggesting that they consist of flux rather than air. It was possible to track the position of individual solder balls in the channel and in Fig. 2 b, areas where the local ordering of particles in chains structures can be seen. This type of random particle chains is known to cause jamming and the fluid has a microstructure resembling that of a glassy state [2]. While the pulsating flow cannot be correlated directly to the formation of these chain structures, the presences of them suggest that jamming as a result of periods with extensive ordering is possible. During periods where the paste is less ordered, flow is less restricted.
Pressurization of the paste using 2 MPa resulted in a large, permanent, rearrangement of the solder paste particles, Fig. 3. The interface of the solder paste balls moved about 1.6 mm into the channel, corresponding to a volume compression of 1.6%. Upon releasing the pressure, 0.1% of the compression was showed to be elastic and went back after reducing the pressure to 0.1 MPa.

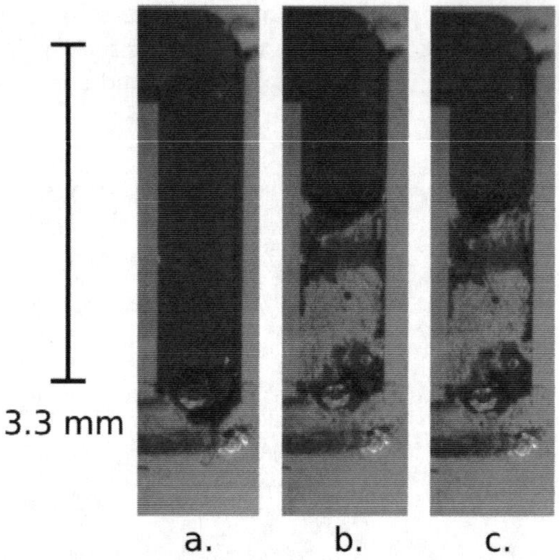

Figure 3. The T-junction of the chip with solder paste and hydraulic oil. In (a), hydraulic oil has been flowed from right to left in the lower channel, cutting the solder paste plug off and forming and interface. In (b), the channel is pressurized from 0.1 MPa to 2 MPa resulting of the solder paste interface moving about 1.6 mm in the channel. After the pressure is returned to 0.1 MPa (c), the interface retracts back to the T-junction by less than 0.1 mm.

The low optical contrast between the flux in the solder paste and the hydraulic oil used as the pressurizing medium made the meniscus between them not visible; making it difficult to judge to what degree the rearrangement of the solder particles was accompanied with a phase separation of the flux. To make more precise measurements of the compressibility, it is necessary to have better contrast between the flux and the media used for pressurization. This can either be done using a pressurizing fluid that has a different color, or by adding a dye that can give contrast at some wavelength.

For future studies, it would be advantageous if both the inlet and outlet part of the solder paste plug could be monitored at the same time by designing a chip where they are in close proximity to each other. Also, having a smaller channel diameter of the T-junction, where the length difference is detected, compared to the meander containing the bulk of the paste would make it possible to detect even smaller changes in compressibility. Further, a smaller diameter of the T-junction should also allow for a more focused cut off of the solder paste segment, thereby further increasing the resolution of the method.

As shown by the initial studies presented here, several complex behaviors are involved when high solid content fluids flow through microchannels. With the use of transparent, pressure tolerant glass chips, insight into the fluid handling can be given.

4. Conclusion

- The transparent microfluidic high-pressure glass chip with a special developed interface was shown to be suitable for flow studies of solder paste with a high solid content.
- The chip was pressure tolerant up to 17.3 MPa making it possible to study effects of applying pressure to the fluid. Operation with the low restriction fluid interface was demonstrated up to 6 MPa.
- The flow rate through the chip displayed a pulsing behavior that could be connected to local ordering effects and jamming of particles in the solder paste.

- Pressurization with 2 MPa resulted in a permanent re-organization of the solder paste balls, corresponding to a compaction of 1.6%. Only 0.1% of the compression was elastic and could be retrieved as the pressure was released.

Acknowledgment

Funding for the laboratory facilities by Knut and Alice Wallenberg Foundation is acknowledged.

References

[1] K. Jensen, B. J. Reizman and S. G. Newman, 2014, *Lab Chip*, **14**, pp. 3206-3212
[2] S. M. Peker, and S. S. Helvaci, 2011, Solid-Liquid Two Phase Flow, *Elsevier*
[3] Y. W. Kim and J. Y. Yoo, 2012, *Opt Lasers Eng,* **50**, pp. 87-98
[4] X. Xiangchun, J. Zhu, and C. Church, 2010, *Microfluid Nanofluidics,* **9**, pp. 1-16
[5] J. Nilsson, M. Evander, B. Hammarström, and T. Laurell, 2009, *Anal. Chim. Acta*, 649-2, pp. 141-157
[6] Z. Wang and J. Zhe, 2011, *Lab Chip*, **11**, pp. 1280-1285
[7] A. Pietrikova and M. Kravcik, 2012, *35th International Spring Seminar on Electronics Technology (ISSE)*, Bad Aussee **9** May, pp. 138-143
[8] F. Tu, Y. Wang, J. Zhou, and F. Xue, 2016, *17th International Conference on Electronic Packaging Technology (ICEPT),* Wuhan, **16** . pp. 1112-1116
[9] S. Gu, X. Jiao, J. Liu, Z. Yang and Q. Jiang, 2016, *Micromachines,* **112**
[10] N. Coenen, 2013, *ICSR (Soldering and Reliability) Conference Proceedings,* May, 14
[11] M. Andersson, K. Hjort and L. Klintberg, 2016, *J. Micromech. Microeng.*, **26**, 095009

IOP Conf. Series: Journal of Physics: Conf. Series **922** (2017) 012011 doi:10.1088/1742-6596/922/1/012011

Ionic liquid electrolyte for supercapacitor with high temperature compatibility

Mazharul Haque, Qi Li, Volodymyr Kuzmenko, Anderson D. Smith, Peter Enoksson

Chalmers University of Technology, Göteborg, Sweden

mhaque@chalmers.se

Abstract. This work describes the electrochemical investigation of two ionic liquids (ILs), 1-ethyl-3-methylimidazolium acetate (EMIM Ac) and 1-butyl-3-methylimidazolium chloride (BMIM Cl), as electrolytes in supercapacitors (SC). A comprehensive study on high temperature (HT) endurance that is required for system integration in microelectronics has also been carried out. It has been found that EMIM Ac containing SC performs better than a BMIM Cl containing SC, and HT treatment improves the capacitive performance.

1. Introduction

Supercapacitors are getting tremendous attention in recent years as they can fill the gap between traditional capacitors and batteries with their high power capability, decent energy density and remarkable lifetime [1]. Owing to their unique performance characteristics, SCs are often integrated in microelectronics systems [2]. Many concurrent encapsulation and packaging techniques for printed circuit board (PCB) electronics, such as reflow soldering [3], are high temperature (HT) processes, which require the SC to withstand elevated temperature (more than 100 °C) for a certain period of time [2]. The HT exposure specifically requires an electrolyte that is compatible with harsh temperature and pressure. Aqueous and organic electrolytes are unsuitable for HT exposure due to the presence of water or high vapor pressure [4]. In contrast, Ionic liquids (ILs), also known as room temperature (RT) molten salts are an appropriate option as an electrolyte because of their high thermal stability [5]. Moreover, they possess valuable intrinsic properties such as low vapor pressure, non-flammability, high electrochemical stability and wide voltage range [5].

In this work, we have evaluated two different ILs, 1-ethyl-3-methylimidazolium acetate (EMIM Ac) and 1-butyl-3-methylimidazolium chloride (BMIM Cl), as potential electrolytes for SCs that can withstand the HT required for microelectronics encapsulation. SC devices were fabricated by using activated carbon (AC) electrodes, EMIM Ac or BMIM Cl electrolyte and glass fiber (GF) separators. In order to mimic the packaging and encapsulation requirements, fabricated devices were exposed to HT and compared in terms of their electrochemical performance.

2. Experimental

2.1. Electrode preparation

For electrode preparation, initially AC powder and carbon black were mixed and finely ground in a quartz mortar. Afterwards the mixture was added to a beaker containing acetone and polytetrafluoroethylene (PTFE) binder (in the form of 60 wt.% dispersion in H_2O suspension). The weight ratio of the AC powder, carbon black and PTFE was 80:10:10. The mixture in acetone was kept in a sonicator for 30 min and then transferred to a 75 °C water bath. After the solvent (acetone) evaporation, an obtained dough was processed into a freestanding thin film by a rolling press machine. Then, the film was dried in vacuum for 24 h at 80 °C. Pellets with 10 mm diameter were punched out from the dried film and used as electrodes. A single electrode was approximately 3 mg in mass, 100 µm in thickness, and 10 mm in diameter.

2.2. SC device assembly and high temperature exposure
A symmetrical two electrode SC device was prepared by sandwiching two AC electrodes with a separator/electrolyte in between and assembled in a Swagelok® cell. In total, four devices were assembled, three with EMIM Ac electrolyte and one with BMIM Cl electrolyte. Two devices containing EMIM Ac and BMIM Cl electrolytes were compared without any HT treatment. The other two devices containing EMIM Ac were subjected to HT treatment at 190 °C for 2 h and at 120 °C for 6 h. HT treatment was provided by placing the cells into DZF-6020 oven.

2.3. Electrochemical measurement
All the devices were kept 'as is' for 24 h before measurement in order to give enough time for the electrolyte ions to penetrate into the electrodes. The following electrochemical measurements were conducted on a Gamry Reference 3000 AE potentiostat to evaluate the electrochemical performance of the devices: cyclic voltammetry (CV), galvanostatic charge discharge (GCD), cyclic charge/discharge (CCD) and electrochemical impedance spectroscopy (EIS). Capacitance was calculated from both CV and GCD measurements according to the following equations:

$$C_s = \frac{4}{m\Delta V v_s}\int_{V_1}^{V_2} I(V)dV \qquad \text{(F/g)} \qquad (1)$$

$$C_s = \frac{4 \times I_d \times \Delta t}{m \times \Delta V} \qquad \text{(F/g)} \qquad (2)$$

where C_s is the specific capacitance of a single electrode, $I(V)$ is the current response, $V_1$and V_2 are lower and upper potential limits, respectively, ΔV is the working voltage range excluding IR drop, v_s is the scan rate, I_d is the discharge current, Δt is the discharge time and m is the total mass of both electrodes.
Power density and energy density of a device were calculated from GCD measurements based on the following equations:

$$C = \frac{C_s}{4} \qquad \text{(F/g)} \qquad (3)$$

$$E = \frac{C\Delta V^2}{2 \times 3.6} \qquad \text{(Wh/kg)} \qquad (4)$$

$$P = \frac{E \times 3600}{t} \qquad \text{(W/kg)} \qquad (5)$$

where C is the specific capacitance of the device, ΔV is the voltage range excluding IR drop, E is the energy density of the device and P is the power density of the device.

3. Results and discussions

3.1. Performance comparison of non-treated device
Figure 1 presents the CVs of EMIM Ac and BMIM Cl containing SCs at different scan rates. At a fast scan rate of 200 mV/s, EMIM Ac containing cell exhibits a fairly symmetric and semi rectangular CV curve (Figure 1a). The deviation from an ideal rectangular shape (a characteristic of electrochemical

double layer mechanism, i.e. EDL behavior) is due to the high resistive nature of ILs. However, the performance improves at a slow scan rate. The specific capacitance of EMIM Ac containing SC at 5 mV/s is calculated to be of 82 F/g which is almost 2.5 times more than specific capacitance at 200 mV/s.

For the BMIM Cl containing SC, the CV curve (Figure 1b) at a fast scan rate of 200 mV/s is less rectangular and rather asymmetric. This suggests a presence of very high resistance in the device, which might originate from the higher viscosity of BMIM Cl compared to the viscosity of EMIM Ac electrolyte. Also, at slow scan rates, there is a noticeable deformation in the end of charging and in the beginning of discharging cycles. A plausible explanation for the deformation can be the higher reactivity (corrosiveness) of Cl^- anion in BMIM Cl compared to Ac^- anion in EMIM Ac [6, 7], which facilitates faradic reactions. The specific capacitance is calculated to be 75 F/g at the slowest scan rate, while it decreases to 31 F/g at a high scan rate of 200 mV/s.

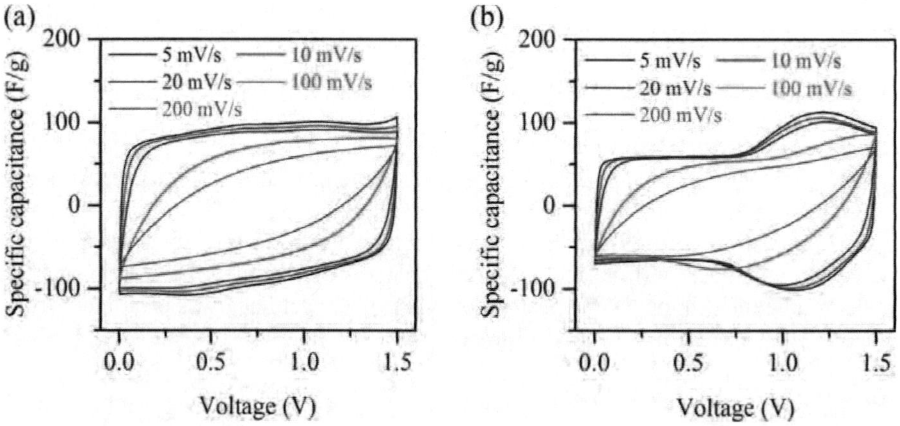

Figure 1. CVs of SC containing EMIM Ac (a), BMIM Cl (b). The total mass of electrodes for EMIM Ac and BMIM Cl containing SC are 5.57 and 6.26 mg, respectively.

Figure 2a and Figure 2b demonstrate the GCD curves of EMIM Ac and BMIM Cl containing SCs. Voltage loss, also known as IR drop, is noticeable even at a low current density of 0.5 A/g for both devices due to slow kinetic properties of ILs at RT. The typical triangular shape of EMIM Ac containing SC (Figure 2a) at all current densities represents almost perfect EDL behavior. Non-linear charge and discharge curves of BMIM Cl containing SC clearly indicates the presence of faradaic reactions, which is in agreement with the CV results.

Figure 2c represents the CCD curve of both devices. Interestingly, capacitance gradually increases to 104% at 5000th cycle for EMIM Ac containing SC. This unusual increase of capacitance may has been caused by the continuous wetting of electrode pellets by the electrolyte. Also repetitive charge/discharge might have opened up more pores of AC electrodes. In a sharp contrast, capacitance decreased dramatically over the measurement on BMIM Cl containing SC and retained only 86% at 3500th cycle. The inferior stability of BMIM Cl compared to EMIM Ac should be originated from the stronger chemical reactivity of Cl^- anions compared to Ac^- anions, which increases with cycling. Acetate-based ILs are less toxic and corrosive compared to halogen-containing ILs [6, 7].

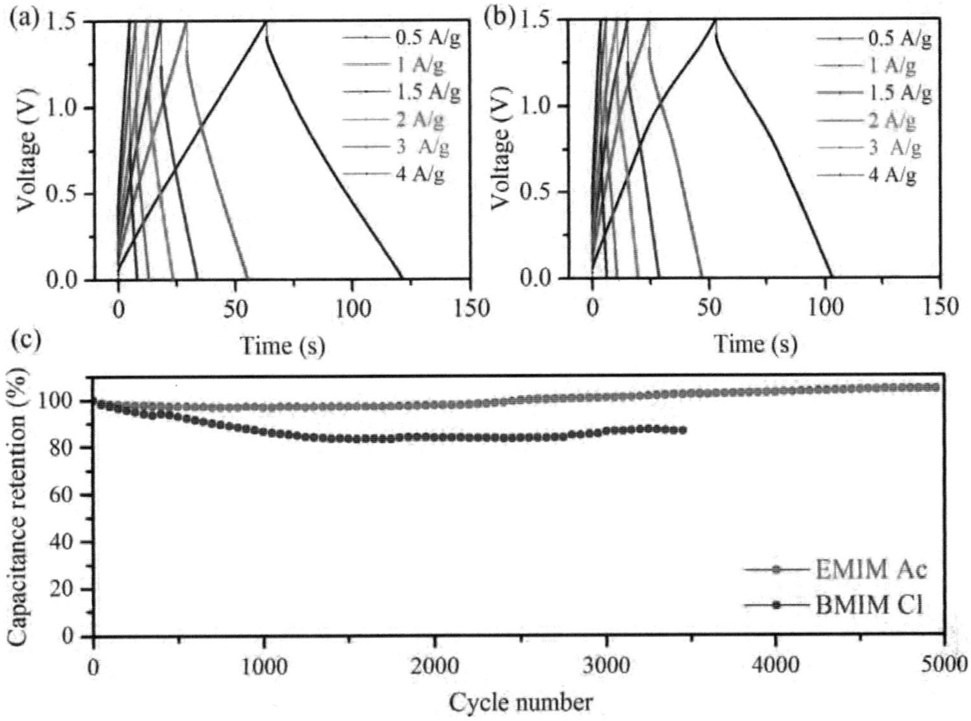

Figure 2. GCD curves at selected current densities of EMIM Ac (a), BMIM Cl (b) containing SC, and CCD measurement at 3 A/g (c).

Comprehensive comparison of performance metrics is summarized in Table 1. Based on the results, it can be concluded that the EMIM Ac containing SC demonstrates superior EDL behavior compared to BMIM Cl containing SC. This can be due to the better compatibility of EMIM Ac electrolyte with AC electrode. Also IL with shorter alkyl chain usually has lower viscosity compared to IL with longer alkyl chain [8]. In general, viscosity is inversely proportional to ionic conductivity, therefore, EMIM Ac with the shorter ethyl chain possesses comparatively higher conductivity than BMIM Cl with the longer butyl chain, which results in better capacitive behavior for the EMIM AC containing SC. Cyclic stability measurements clearly indicate that the EMIM Ac containing SC holds greater stability as an electrolyte compared to BMIM Cl.

3.2 Non-treated vs heat-treated device

Considering the possible requirements on packaging and encapsulation of a SC device in electronics systems, electrochemical measurements after HT treatment were carried out and results are presented in this section. Based on the results of the non-treated devices, only EMIM Ac containing SCs were selected and subjected to heat treatment. After HT treatment, devices were cooled down to RT before starting electrochemical measurements. The differences between treated devices and a non-treated device are observed in order to investigate the thermal influence on the electrochemical performance.

Figure 3a presents CV curves of HT treated and non-treated devices containing EMIM Ac electrolyte. The curves not only maintain standard rectangular shape but also become wider with HT treatment. This phenomenon indicates improved charge accumulation on the electrode/electrolyte interface during the heat treatment. The calculated specific capacitances of the SCs without and with treatment at 120 °C for 6 h and 190 °C for 2 h are 73, 79 and 81 F/g, respectively. Figure 3b demonstrates the GCD curves of the same devices. It can be seen that the duration of charge and discharge time increases with HT treatment. Also, the IR drop reduces significantly suggesting improved kinetics of the electrolyte ions. The calculated specific capacitance of the corresponding three SCs are 71, 78 and 77 F/g. The slightly different values of the specific capacitance calculated from CV and GCD measurements are expected as the applied currents in both techniques were different. Overall, the results clearly indicate that HT treatment improves the capacitive performance. This is mostly due to an accelerated electrolyte ion penetration into an electrode material as a consequence of HT exposure.

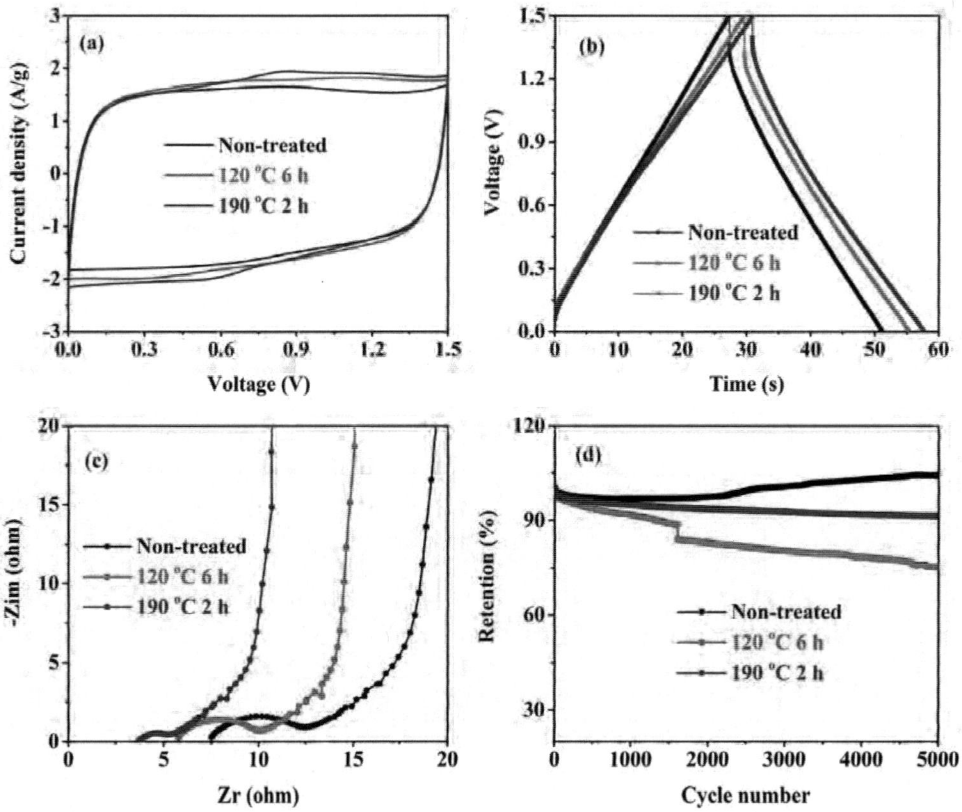

Figure 3. Comparison of electrochemical performance of EMIM Ac containing SC with and without temperature treatment: CVs at 20 mV/s (a), GCD plots at 1 A/g (b), Nyquist plots (c), CCD at 3 A/g (d). The total mass of electrodes for non-treated, 120 °C and 190 °C treated cells are 5.57, 5.55 and 7.61 mg, respectively.

Figure 3c represents Nyquist plots instigated from EIS measurements. The initial intercept of a plot in the x-axis at high frequency region represents bulk electrolyte resistance. It can be clearly seen that the resistance decreases with the heat treatment. The equivalent series resistance (ESR) is obtained from the semicircle in the mid-high frequency region and it is also decreased and shifted to the left. Calculated ESR values are 12.68, 10.33 and 5.39 ohms for non-treated, treated at 120 °C and treated at 190 °C cells, respectively.

Figure 3d compares the cycling stability of the non-treated and treated devices. The capacitance retention of 120 °C treated cell is not up to the mark as it shows a 23% loss in the end of 5000 cycles. The 190 °C treated cell exhibits better capacitance retention than the 120 °C treated one, retaining 94% of the initial capacitance. When treated at lower temperature (120 °C), the cell was exposed for longer duration (6 h) than treated at 190 °C (2 h). Therefore, it is likely that duration might have played more crucial role than temperature itself. As it mentioned before, unlike the treated cells, the non-treated cell exhibits increased capacitance over the cycling period.

Table 1 Electrochemical properties of SCs containing different electrolytes [a]

Type	Electrolyte	Specific capacitance (F/g)	Energy density (Wh/kg)	Power density (W/kg)	Capacitance retention (%)
Non-treated	EMIM Ac	82	4.52	680	104
	BMIM Cl	75	4.11	665	86
Heat treated	EMIM Ac/120	90	4.74	663	77
	EMIM Ac/190	93	5.27	703	94

[a] *Specific capacitance is calculated from CV at 5 mV/s, energy density and power density are calculated from GCD at 1 A/g, and capacitance retention is relative to first cycle value at 3 A/g from CCD measurement.*

4. Conclusion and future work

A thorough electrochemical investigation of EMIM Ac and BMIM Cl ILs as potential electrolytes for SCs has been carried out. EMIM Ac outperforms BMIM Cl by means of capacitance and cyclic stability. Moreover, HT treatment (at 190 °C for 2 h) of SC containing EMIM AC not only retained the capacitive behavior but also improved the performance, which assures the feasibility of integrating the SC in microelectronics that requires elevated temperature treatment. Compared to higher temperature, longer duration of HT treatment is more detrimental to cycling stability. As a future work, more detailed investigation should be focused on understanding the influence of heating duration on cyclic stability. Lastly, in depth electrochemical investigation of EMIM AC electrolyte containing SC operating at different elevated temperatures and finding the temperature limit should be beneficial as well.

Acknowledgements

This work was funded by the European Union's Horizon 2020 research and innovation program [grant number 644378, smart-MEMPHIS project], and Vinnova UDI project Energy Harvesting Toolkit.

References

[1] P. Simon, Y. Gogotsi, Materials for electrochemical capacitors, Nat. Mater., 7 (2008) 845-854.
[2] Y. Wang, W. Lai, Z. Jiang, C. Yang, All-printed paper based surface mountable supercapacitors, IEEE Transactions on Dielectrics and Electrical Insulation, 24 (2017) 676-681.
[3] M. Brachet, D. Gaboriau, P. Gentile, S. Fantini, G. Bidan, S. Sadki, T. Brousse, J. Le Bideau, Solder-reflow resistant solid-state micro-supercapacitors based on ionogels, Journal of Materials Chemistry A, 4 (2016) 11835-11843.
[4] G. Xiong, A. Kundu, T.S. Fisher, Thermal effects in supercapacitors, SpringerBriefs in Applied Sciences and Technology (2015).
[5] M. Galiński, A. Lewandowski, I. Stępniak, Ionic liquids as electrolytes, Electrochim. Acta, 51 (2006) 5567-5580.
[6] C.P. Fredlake, J.M. Crosthwaite, D.G. Hert, S.N. Aki, J.F. Brennecke, Thermophysical properties of imidazolium-based ionic liquids, J. Chem. Eng. Data, 49 (2004) 954-964.
[7] R.C. Remsing, R.P. Swatloski, R.D. Rogers, G. Moyna, Mechanism of cellulose dissolution in the ionic liquid 1-n-butyl-3-methylimidazolium chloride: a 13 C and 35/37 Cl NMR relaxation study on model systems, Chem. Commun. (Cambridge, U. K.), (2006) 1271-1273.
[8] S. Fendt, S. Padmanabhan, H.W. Blanch, J.M. Prausnitz, Viscosities of acetate or chloride-based ionic liquids and some of their mixtures with water or other common solvents, J. Chem. Eng. Data, 56 (2010) 31-34.

IOP Conf. Series: Journal of Physics: Conf. Series **922** (2017) 012012 doi:10.1088/1742-6596/922/1/012012

Miniaturization of Compensational Spring System using the Highly Nonlinear Curved beams in MEMS

Bogdan Vysotskyi[1,2], Denis Aubry[2], Philippe Gaucher[2], Elie Lefeuvre[1], Xavier Le Roux[1] and Fabien Parrain[1]

[1] Centre de Nanosciences et Nanotechnologies, CNRS, Université Paris-Sud, Université Paris-Saclay, C2N-Orsay, 91405, France

[2] MSSMat, CentraleSupélec, Université Paris-Saclay, Châtenay-Malabry, CNRS UMR 8579, 92295, France

bogdan.vysotskyi@u-psud.fr

Abstract. In this work an approach to minimize a non-linear system with a flat region in the force-displacement characteristics is studied. Design, simulation, fabrication and characterization of the initial system with the linear compensational spring and that of a miniaturized one without it are presented. The presented highly nonlinear curved beam system shows the same behavior as that with compensational spring system in terms of force-displacement both in the finite element simulation and in the experiment. Thus, an equivalent miniaturized system is proposed with the surface footprint reduced by 2.38 times only by designing means, without changing the technological steps.

1. Introduction

The problematics of miniaturization of separate mechanical elements in micro-electro-mechanical systems (MEMS) remains actual [1]. There are two main techniques that are can be referred in miniaturization of the MEMS devices: either the straightforward decrease in the size of elementary components [2] (which is limited by the available technology of fabrication and design rules, mainly related to anisotropic etching), or by the design of alternative structural elements shapes [3] (general design methodology is not defined in this approach).

Springs, as a structural element in MEMS, are considered to be of crucial importance, as far as they are used in an extremely wide variety of devices [4,5]. Typically, non-linear springs are widely used in a variety of energy harvesting applications [6], resonators [7], switches and actuators [8].

In our research, a need for miniaturized nonlinear springs with a flat force-displacement region has arisen. In the present work, a design, simulation, fabrication and experimental verification of a miniaturized spring is presented in comparison to the original non-miniaturized system.

2. Design of the Double Curved Beams system

A schematics of the compensational springs system and of the highly nonlinear curved beam springs system is shown on the Fig. 1. The compensational spring system (studied in our previous work [9]) is composed by a bistable double curved beam mechanism [10] with attached linear springs with a stiffness of k=21N/m chosen to obtain a flat region in the force-displacement curve. The highly nonlinear curved beam shape was obtained by varying the curvature of the original bistable beam. All considered beams have width of 20μm and thickness of 85μm.

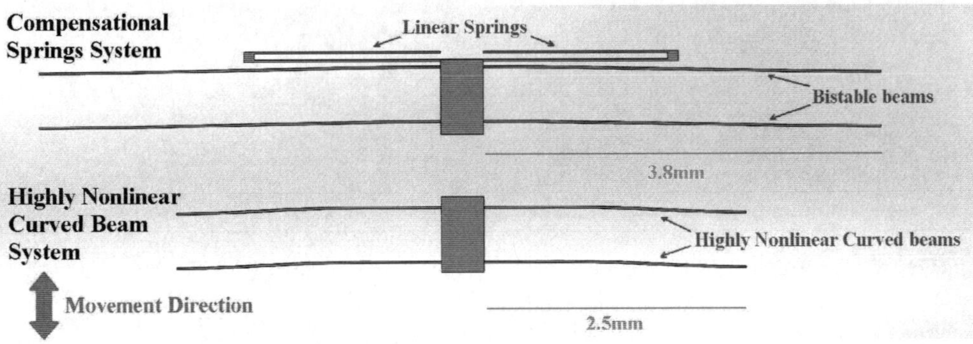

Figure 1. Schematic representation of the original Compensational Springs system and of the miniaturized Highly Nonlinear Curved beams system.

The useful footprint of the layout of the proposed miniaturized system is 0.73mm² versus 1.735mm² of the initial system, which yields a miniaturization of almost 2.38 times in surface using only design means.

3. Simulation

To simulate the mechanical properties of the compensational spring system and its high nonlinearity, a finite element model (FEM) approach had been used. ANSYS models are created for each system, and a static analysis of force-displacement is performed. As it is depicted on the Fig. 2, the force-displacement curves of the two systems fit almost perfectly.

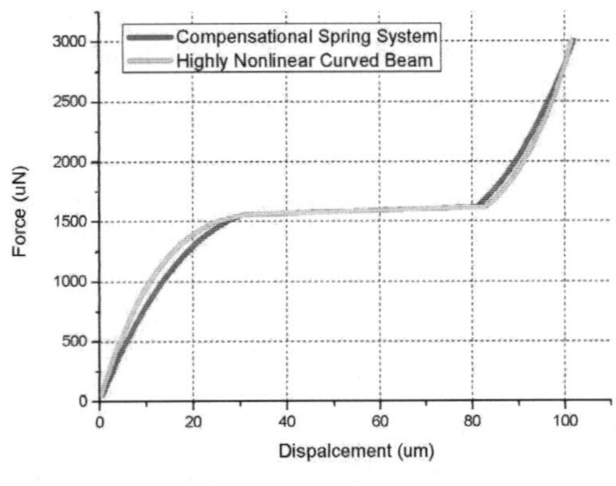

Figure 2. The force-displacement curves of Compensational Spring System and Highly Nonlinear Curved Beam system obtained with FEM simulation.

There are 3 regions present on the presented force-displacement curve: 1 : progressive softening, 2: large compliance (plateau) and 3: progressive hardening [10,11]. The behavior on the first and last regions is the typical for a bistable mechanism, whereas the central nearly-flat region comes from the modification of initial nonlinearity.

4. Experimental part

The systems are fabricated using SOI wafer of 85μm device Si thickness, with Bosch process of Deep Reactive Ion Etching (DRIE). The schematics of the process of fabrication is shown on the Fig. 3. First, a photoresist film is deposited by spin-coating (Fig. 3, A), which is followed by UV lithography and development (Fig. 3, B). Then the device pattern is etched using DRIE of Si device layer (Fig. 3, C). At the last step, the system is liberated with HF vapor etching of the buried oxide layer and the photoresist is stripped off (Fig. 3, D). The optical images of the fabricated spring systems is shown on the Fig. 4.

IOP Conf. Series: Journal of Physics: Conf. Series **922** (2017) 012012 doi:10.1088/1742-6596/922/1/012012

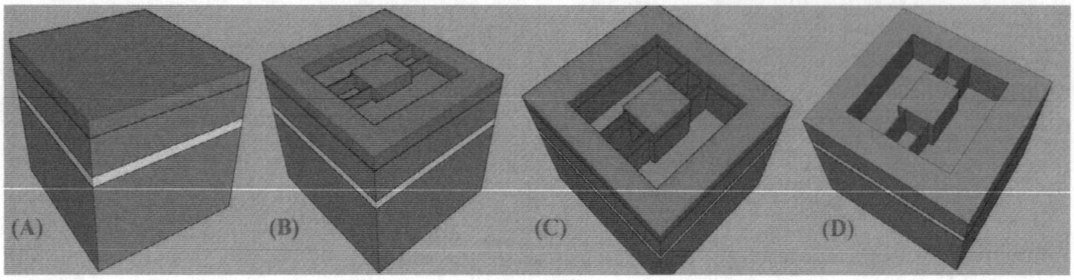

Figure 3. Schematics of the fabrication process. Grey color corresponds for silicon, light grey – for silica, and rose for photoresist.

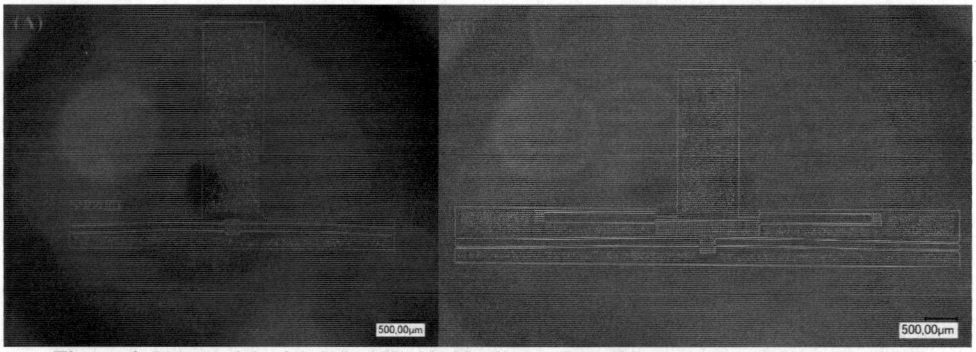

Figure 4. Photos of the fabricated Highly Nonlinear Curved Beam system (A) and Compensational Springs system (B). Note the difference in the footprint size.

To characterize experimentally the fabricated systems a micromechanical testing instrument (FT-MTA02, FemtoTools [12]) is used. It consists of an actuation unit with a micro force sensing probe needle and a microscope that allows aligning the probe with the tested microstructure. Photo of experimental set-up is shown on the Fig. 5, (A). Mechanical reaction force of the studied spring systems is measured with force sensing probe with a 30° inclination of the probe. This angle has been considered in the post-treatment of the measurement results, which are demonstrated on the Fig. 5, (B).

5. Discussions

The measured force-displacement characteristics of both systems are relatively close, and the slight difference is coming from the limitations of the used fabrication techniques. Experimental curves have a shape that is similar to that obtained by FEM. However, the measured reaction force values are almost two times smaller than the ones predicted by simulation, which is easily explained by the roughness and notching present on the sides of the beams due to DRIE process imperfections. Side walls of the springs are observed to be not perfectly vertical due to the limitations of DRIE process. Equivalent (mean) beam thicknesses have been estimated as 17.5µm instead of the 20µm designed.

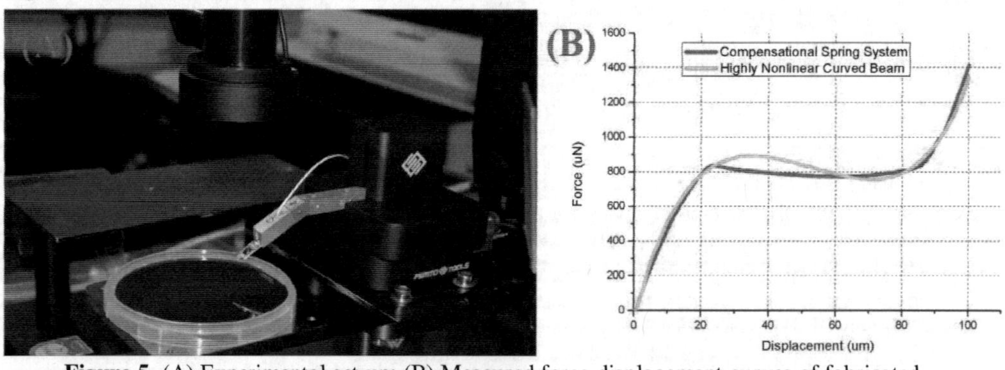

Figure 5. (A) Experimental set-up; (B) Measured force-displacement curves of fabricated Compensational Spring System and Highly Nonlinear Curved Beam system.

6. Conclusion

In the present article, the design, simulation, fabrication and characterization of a highly nonlinear spring system with an almost flat force-displacement region is presented. This system is shown to be mechanically equivalent to compensational spring system, which consists of the combination of two pairs of bistable curved beams and linear spring. However, the footprint on the mask is decreased significantly (2.38 times).
The finite element model is built for both the initial (the one with compensational spring) system and the miniaturized highly nonlinear curved beams system. Both systems are fabricated in silicon using a classical microfabrication technologies. Characterization is performed using a micromechanical testing instrument with a force-displacement sensing probe. The measured force-displacement characteristics for both initial and miniaturized systems are close to each other, and similar to that predicted by FEM analysis. The difference in absolute values between simulations and measurements are coming from the imperfections of the fabrication. Due to the all presented results, it can be said that the initial compensational springs system is successfully miniaturized by using only highly nonlinear curved beams.

Acknowledgements

This work is supported by the "IDI 2015" project funded by the IDEX Paris-Saclay, ANR-11-IDEX-0003-02 and has benefited from the financial support of the LabeX LaSIPS (ANR-10-LABX-0040-LaSIPS) managed by the French National Research Agency under the "Investissements d'avenir" program (n°ANR-11-IDEX-0003-02).

References

[1] M. Tanaka, "An industrial and applied review of new MEMS devices features", Microelectronic Engineering, 84, 1341–1344, 2007;
[2] M. J. Madou, "Fundamentals of Microfabrication: The Science of Miniaturization", 2nd ed., CRC Press, 2002;
[3] C.M.N. Brigante, N. Abbate, A. Basile, A.C. Faulisi, and S. Sessa, "Towards Miniaturization of a MEMS-Based Wearable Motion Capture System", IEEE Transactions On Industrial Electronics, Vol. 58, No. 8, 2011;
[4] R. Liu, H. Wang, X. Li, J. Tang, S. Mao, and G. Ding, "Analysis, simulation and fabrication of MEMS springs for a micro-tensile system", Journal Of Micromechanics And Microengineering 19, 2008;
[5] W.M. Zhang, H. Yan, Z.K. Peng, G. Meng, "Electrostatic pull-in instability in MEMS/NEMS: A review", Sensors and Actuators A 214, 187–218, 2014;
[6] M.F. Daqaq, R. Masana, A. Erturk, and D.D. Quinn, "On the Role of Nonlinearities in Vibratory Energy Harvesting: A Critical Review and Discussion", Applied Mechanics Reviews 66, 2014;
[7] L.L. Li, P.M. Polunin, S. Dou, O. Shoshani, B.S. Strachan, J.S. Jensen, S.W. Shaw, and K.L. Turner, "Tailoring the nonlinear response of MEMS resonators using shape optimization", Applied Physics Letters 110, 2017;
[8] R. Legtenberg, A.W. Groeneveld, and M. Elwenspoek, "Comb-drive actuators for large displacements", Journal Of Micromechanics And Microengineering 6, 1996;
[9] B. Vysotskyi, F. Parrain, D. Aubry, P. Gaucher and E. Lefeuvre, "Innovative Energy Harvester Design Using Bistable Mechanism With Compensational Springs In Gravity Field", Journal of Physics: Conference Series, 773, 2016;
[10] J. Qiu, J. H. Lang, A. H. Slocum, "A Curved-Beam Bistable Mechanism", Journal of Microelectromechanical Systems, 13, No. 2, 2004;
[11] P. Cazottes, A. Fernandes, J. Pouget, M. Hafez, "Bistable Buckled Beam: Modeling of Actuating Force and Experimental Validations", Journal of Mechanical Design 131, 2009;
[12] FemtoTools FT-MTA02 product brochure http://www.femtotools.com/fileadmin/pdf/Product_Brochures/FT-MTA02_Broschure.pdf

IOP Conf. Series: Journal of Physics: Conf. Series **922** (2017) 012013 doi:10.1088/1742-6596/922/1/012013

Prediction of dynamic behaviour of Pt/SiNx suspended beams for their use in harsh environment

Baudry Florian[1], Alain Giani[1], Philippe Combette[1]

[1] IES, Univ Montpellier, CNRS, Montpellier, France

Abstract. This paper reports a mechanical study of double clamped micro beams of an Accelerometerbased on Thermal Convection (ATC). One of the prime application of ATC is shock measurement where the magnitude of acceleration could exceed 20000g [1-2]. The ATC is formed of 3 suspended resistors. The ATC would support a large frequency spectrum of high magnitude due to a shock application. The eigen frequencies has to be found because the resonant effect could destroy the suspended resistors. Analytic results describe the heater resistor as tensile string [4]. Finite element method (FEM) and measurement of resonant frequencies with interferometric microscope validate the tensile string model [5]. Also by experimental measurement, it is shown that the increase of temperature reduces the tensile stress of the beam and then the values of resonance frequencies.

1. Introduction

Thermal micro machined accelerometers have been intensively studied because of their high shock reliability. This reliability is due to lack of seismic mass [1-2]. The main application of this kind of accelerometer is high magnitude shock detection in automotive industry like air bag crash detection or defence applications [3]. The goal of this paper is to demonstrate that the behaviour of the suspended resistors is the same as tensile string [5]. This comparison is done by FEM and by experiments with an interferometric microscope. The innovation of this paper consists in the measurements of resonant frequencies by heating the thermal resistors. The final objective is to study the influence of temperature on resonance frequencies.

2. Physical principle and microstructure design

The ATC is based on free convection in a sealed chamber containing N_2 [1-3]. As shown in figure 1, The heater resistor is placed between two detectors. When the heating resistor is powered, it creates a symmetrical temperature profile. Consequently, the detectors are at the same temperature (continuous line). Although in presence of acceleration, the thermal gradient tends to shift in the same direction than the acceleration (dotted line). By knowing the coefficient of thermal resistance (CTR) and the resistance at 0 °C (R_0) of the platinum resistor, the difference of electric voltage collected between the two detection resistors is converted into difference of temperature. The shift in temperature is proportional to the acceleration [2].

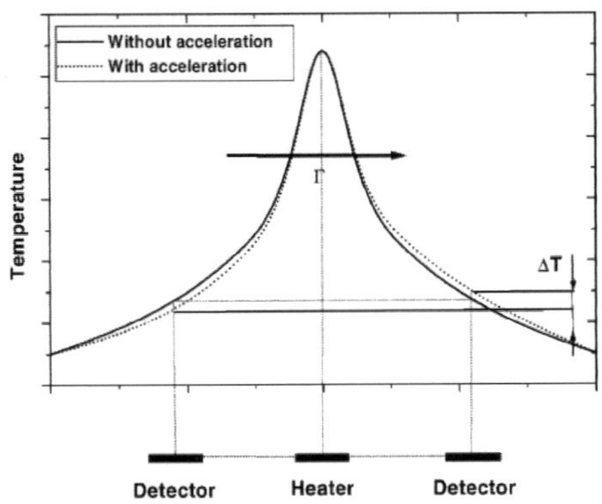

Figure 1: View of the thermal gradient with the disposition of resistors

The heater resistance temperature could exceed 150 °C, then it's important to study the influence of temperature on resonant frequencies.

3. Analytical results and FEM modelisation

The beam is described by a clamped-clamped beam model, with one degree of freedom, given by Euler Bernouilli equation [5-6]. Although in the case of high tensile constrain (L>>h where L the length and h is the thickness of the heater) the resonance frequencies are a first mode integer multiple. The equation below (1) describes the frequency of the nth mode with n=1,2,3 as a function of tensile stress σ, the mass density ρ and the length of the beam L:

$$f_n = \frac{n}{2L}\sqrt{\frac{\sigma}{\rho}}, \quad (1)$$

This model is suitable for beams composed by only one material. Although FEM need to be performed in order to determine the resonance frequencies and the mode shape for a SiNx/Pt beam. The FEM is achieved with ANSYS software. Analysis for two different lengths and widths of the beam is performed: the first is 2000 µm by 50 µm and a second is 500 µm by 10 µm. These dimensions correspond to different device design [2]. The thickness of SiNx beams is around 500 nm. SiNx/Pt beams is 500 nm for SiNx layer and 300 nm for Pt layer. The results of FEM are given in Table 1. The results are classified by compositions and dimensions of the beams:

Table 1 : FEM results

Composition	Size(µm)	f1(kHz)	f2 (kHz)	f3 (kHz)
SiNx	500*10	173	348	526
SiNx	2000*50	44	89	134
SiNx/Pt	500*10	160	322	486
SiNx/Pt	2000*50	33	67	78

However, for FEM it's necessary to know the stress values of SiNx and SiNx/Pt beams. The tensile stress σ is found by using the equation (1). Table 2 shows the tensile stress deduced from the measurement of resonant frequencies at T= 23 °C with ρ_m the average volumic mass of the suspended resistor:

IOP Conf. Series: Journal of Physics: Conf. Series **922** (2017) 012013 doi:10.1088/1742-6596/922/1/012013

Table 2 : Tensile stress deduced from equation (1)

Composition	Size(µm)	ρ_m(gr.cm^{-3})	σ (MPa)
SiNx	500*10	3,15	100
SiNx	2000*50	3,15	107
SiNx/Pt	500*10	10	253
SiNx/Pt	2000*50	10	192

The resonance frequencies are a first mode integer multiple of the first mode and the mode shapes are the same as tensile string. This result is valid for every size and composition of the beam studied. The next part presents the experimental values of eigen frequencies and mode shapes found by an interferometric microscope.

4. The Experimental Data and discussions

4.1 Modes shapes and resonant frequencies

Experimental measurement of frequencies and mode shape is done by the *Fogale photomap 3D,* optical profiling system. This system is combined with an ultrasonic transducer which allows scanning bandwidth in order to find eigen frequencies and mode shapes. The mode shape at the resonant frequencies are found by vibrometry in stroboscopic light [7] and are shown in Figure2. The measurements have been done on SiNx/Pt resistor of 500 µm by 10 µm. In order to get a clear view of the flexural displacement, the ratio of Y-axis: X-axis, was chosen around 1:5000:

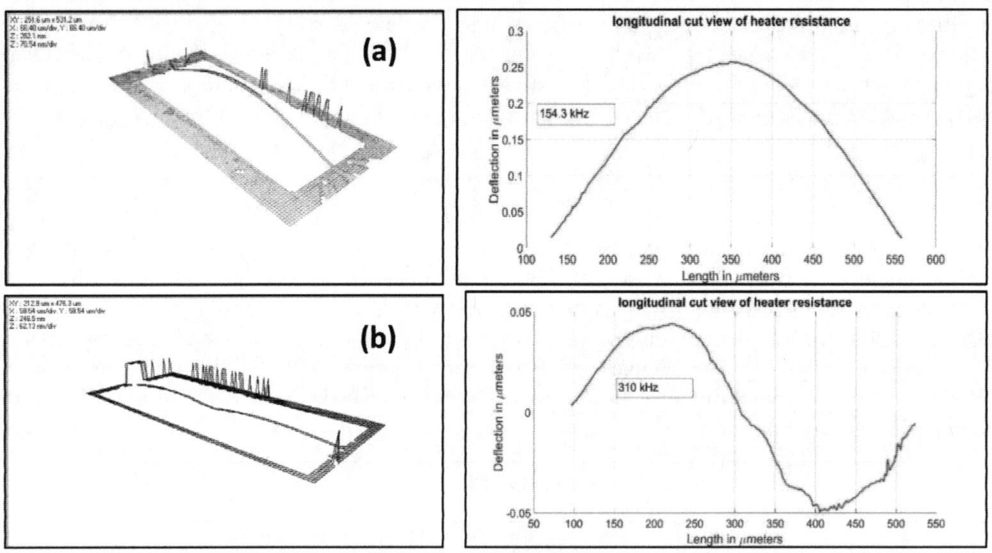

Figure 2: Experimental deformed shape of the (a) first, (b) second mode in 3D view and longitudinal cut view

The behaviour of SiNx and SiNx/Pt beam were the same as tensile string. The values in table 3 shows very good agreement between experimental (Table 3) and FEM results (Table 1):

Table 3: Measured resonance with optical profilometer

Composition	*Size(µm)*	*f1(KHz)*	*f2*	*f3*
SiNx	500*10	173	349	/
SiNx	2000*50	45.80	92.02	139.4
SiNx/Pt	500*10	154.06	309.6	468.3
SiNx/Pt	2000*50	34.46	68.91	103.48

FEM shows a good prediction of the resonant frequency and the shape of deformation.

4.2 Influence of the temperature on resonant frequencies

The operating heater resistor temperature was around 200 °C. With a DC generator an electric current through this resistor was set. The knowledge of CTR and Ro allows us deducing temperature as a function of the electric current. The result of SiNx/Pt 500 by 10 µm of beam for five differents temperatures are gathered in table 4. The eigen frequency of the first mode and the equivalent tensile stress with equation (1) were calculated:

Table 4: Eigen frequencies and tensile stress as a function of temperature

T (°C)	*f1*(kHz)	σ *(MPa)*
21	159,2	253
47	154,9	239
78	145,9	212
122	89,3	80
151	51,8	27

The increase in temperature shifts the resonant frequency to lower frequencies. Unfortunately, it is difficult to measure the resonant frequency at a temperature higher than 150 °C, because of mechanical bulking of the beam [5]. In figure 3, f(T)=*f* with equation (2) is fitted, with the young modulus E [MPa], and α the thermal expansion [K^{-1}]:

$$f_n = \frac{n}{2L}\sqrt{\frac{\sigma_0 + E \times \alpha \times (T - T_0)}{\rho}}, \quad (2)$$

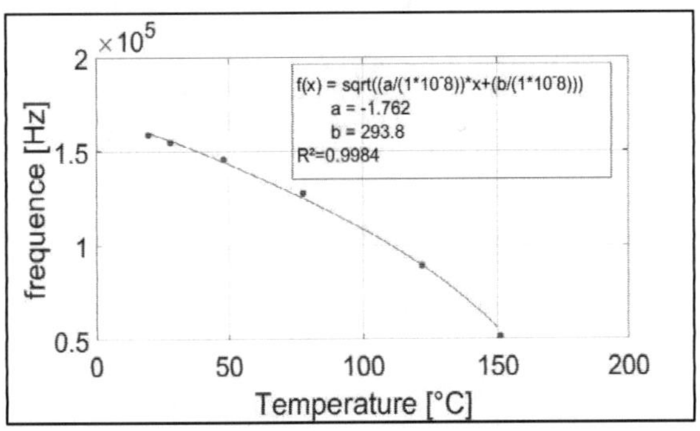

Figure 3: resonant frequencies vs Temperature and its square root regression (2)

Experiment and theoretical model are well matched. By extrapolation for a sample of SiNx/Pt beam of 500 by 10 µm at 200 °C, the resonance frequency is 18,500 kHz.

5. Conclusion

FEM and experiments allows us to conclude that Euler Bernoulli model in the case of high tensile stress is suitable to describe suspended resistor of ATC. By experimental measurements, the result of 500 by 10 µm beams out of plane resonant frequency is around 154 kHz at ambient temperature. Nevertheless, the increase of temperature reduces the value of resonant frequency by reducing the value of tensile stress. The next step of our work is to find a link between the amplitude of acceleration and the deflection of the suspended resistor. In order to estimate the critical acceleration before damaging the device, a combined experimental-analytical approach is possible by measuring the quality factor and resonant frequency [10]. Finally, FEM approach could be carried out to determine the maximum acceleration in function of the frequency.

Acknowledgment
We want to thank the Direction Général de l'Armement (DGA) for the financial support of the thesis of Florian Baudry under n°2016600049 and UMR 5214. We are grateful to the technician and engineer of the "Centre de technologie en nano et microélectronique" of the university of Montpellier and more especially Mr. Jean Lyonnet et Mr. Michel Ramonda for their technical and scientific support.

References

[1] F. Mailly, A.Giani, A. Martinez , R. Bonnot, P. Temple-boyer, A. Boyer 2003, *sensor & Actuator* « Micromachined thermal accelerometer » *vol 103 page 359-353.*

[2] A. Garrault, 2011 *Thesis, University of Montpellier* « Amelioration des performances et nouveau concept de détecteurs de capteur inertiels à détection thermique », France.

[3] A. Garrault, Philippe Combette, Johann Courteaud and Alain Giani 2003, Sensor & Actuators « Effect of the detector width and gas pressure on the frequency response of a micromachined thermal accelerometer » A 103 (2003) 359-363, France.

[4] P. Temple-Boyer, C.Rossi, E. Saint-Etienne and E. Scheid, 1997, *American Vacuum society* « Residual stress in low pressure chemical vapor deposition SiNx films deposited from silane and ammonia », p 2003-2007.

[5] C. Morelle *et al*, *2016*, IEEE "Gallium nitride MEMS resonators: How residual stress impacts design and performances," *2016 Symposium on Design, Test, Integration and Packaging of MEMS/MOEMS (DTIP)*, Budapest, pp. 1-4

[6] Siebe Bouwstra and Bert Geijse, 1991, IEEE « On the resonance frequencies of microbridges », University of twente, Netherland,

[7] A. Bosseboeuf, S. Petitgrand, *SPIE 5145-04,* « Application of microscopic interferometry techniques in the MEMS field », 2003.

[8] Y. –j Su, C. –F. Qian, M. –H. Zhao and T. –Y. Zhang, « Microbridge testing of silicon oxide/silicon Nitride Bilayer Films deposited on silicon wafers », Acta mater 48 (2000) 2843-2857, china 2000

[9] Robin Sh-Islam, june 2012, *Thesis* "Single and double clamped resonator sensors - Design, fabrication and testing", Royal institute of Technology, Sweden.

[10] Subbramanian Sundaram, Maurizio Tormen, Branislav Timotijevic, Robert Lockhart, Thomas Overstolz, Ross P Stanley and Herbert R Shea, 2011 *Journal of micromechanics and Microengineering* "Vibration and shock reliability of MEMS: modelling and experimental validation "21 2011 045022.

IOP Conf. Series: Journal of Physics: Conf. Series **922** (2017) 012014 doi:10.1088/1742-6596/922/1/012014

Graphite paper / carbon nanotube composite: a potential supercapacitor electrode for powering microsystem technology

Qi Li, Anderson D. Smith, Mazharul Haque, Agin Vyas, Volodymyr Kuzmenko, Per Lundgren, Peter Enoksson

Chalmers University of Technology, Gothenburg, Sweden

qili@chalmers.se

Abstract. This work describes fabrication of a flexible supercapacitor electrode. The fabrication starts with graphite paper (GP). Carbon nanotubes (CNTs) are then grown directly to the carbon surface by chemical vapor deposition (CVD), forming a heterogeneous structure of GP/CNT. The integration of CNT enhances capacitive performance while maintaining the flexibility of GP, thus making GP/CNT a promising supercapacitor electrode material for potentially powering microsystem technology.

1. Introduction

Recent advances in MEMS allow nearly seamless integration of an array of actuators and/or sensors to multi-functionalize a device [1]. Benefiting from that, smart systems such as wearable electronics have attracted tremendous attention worldwide and a rapid development is ongoing in recent years to explore the potential of these technologies [2]. Unfortunately, many potential applications are limited by power management [3] and the lifetime of the energy storage component [4].

Supercapacitors, also known as ultracapacitors and electrochemical capacitors, have the ability to deliver near instantaneous energy, work with less complicated circuits and have almost limitless lifetime [5, 6]. These characteristics make this type of energy storage devices suitable for powering MEMS-based systems, especially for those with power consumption as low as microwatts. Additionally, since form factor is critical for wearable technology and many other miniaturized smart systems [1], flexibility of the supercapacitor is advantageous. The flexibility allows the possibility of rolling or folding up an energy storage device to achieve high space-efficiency.

In this paper, we present a fabrication of a flexible electrode material for supercapacitors. The fabrication begins with a sheet of graphite paper (GP) and then compositing with direct carbon nanotube (CNT) growth by chemical vapor deposition (CVD) in 10 seconds. The resulting GP/CNT material is flexible, shows high areal capacitance and cycling stability. Since the GP substrate has a good heat dissipation ability [7], the supercapacitors based on this GP/CNT composite is also possibly capable of carrying high working current without over-heating problems. Taken all into account, GP/CNT composite are potentially suitable for supercapacitors integrable to various advanced electronics.

IOP Conf. Series: Journal of Physics: Conf. Series **922** (2017) 012014 doi:10.1088/1742-6596/922/1/012014

2. Experimental

2.1. Fabrication of GP/CNT

The fabrication of GP/CNT consists of two primary steps: catalyst deposition onto the GP and subsequent CVD growth. First, 2 nm Fe and 3 nm Al_2O_3 layer was deposited successively on the surface of graphite paper (GP) using e-beam evaporation. The arrangement of Fe under Al_2O_3 layer is designed for tip growth of CNTs during CVD process. After catalyst deposition, the GP was placed on a graphite heater mounted in a glass chamber at a pressure below 0.1 mbar. A hydrogen flow of 837 sccm was then introduced into the chamber and the heater ramped up to 500 °C and maintained at this temperature for 3 min. Afterwards, the heater temperature is rapidly increased to 700 °C, while acetylene at a rate of 240 sccm was purged into the chamber for 10 seconds – resulting in vertical CNT growth of 1-2 μm in length.

2.2. Materials characterization

The morphology of GP and GP/CNT materials was examined by scanning electron microscopy (using a SEM, Zeiss Supra 60 VP). Raman spectra and spectra mapping were recorded using a Raman spectrometer (Horiba) with a 638 nm excitation laser.

2.3. Electrochemical measurements

A supercapacitor device (2032 coin cell) was assembled using GP/CNT composite as the electrodes for the initial demonstration of performance. Two identical electrode pellets with diameter of 14 mm were punched out of the corresponding films with a disc cutter. The electrodes were separated by a Whatman® glass fiber membrane of 16 mm diameter. The coin cell was then filled with 50 μl 6 M KOH electrolyte.

The electrochemical measurements were conducted with a Gamry Reference 3000AE at room temperature, 21 °C. The measurements included cyclic voltammetry (CV), galvanostatic charge/discharge (GCD) and cyclic charge/discharge (CCD).

The areal capacitance (mF cm^{-2}) was calculated from GCD measurements using $C = I_{dis}t_{dis} / V$, where I_{dis} (mA cm^{-2}) is the constant discharge areal current (area normalized to 1.5383 cm^2), t_{dis} (s) is the discharge time, and V is the cell voltage excluding IR drop. The areal energy density E (μWh cm^{-2}) and areal power density P (μW cm^{-2}) were obtained by applying the following equations, respectively: $E = 0.5\,C\,V^2 / 3.6$, $P = 3600\,E / t_{dis}$.

3. Results and discussion

3.1 Materials characterization

The surface morphology of GP and GP/CNT is shown in Figure 1. The GP material of grey colour has a relatively smooth surface (Figure 1a). In contrast, the surface becomes rougher since it is covered with CNT bundles (Figure 1b) after the catalyst deposition and CVD process, and the material appearance turns black. A typical structure of the CNT bundles is displayed in Figure 1c-d: CNTs of 1-2 μm attach directly to the carbon (graphite) substrate, while Fe and Al_2O_3 residing at the CNT tips are lifted to the surface. Such a structure is a result of the specific definition of catalyst layers, i.e. Fe under Al_2O_3, which enables an tip growth of CNT [8]. The surface of GP/CNT is non-continuous (Figure 1b-c), because of cracking of the Al_2O_3 layer in the presence of hydrogen gas at elevated temperature. It is worth noting that the graphite substrate has great flexibility (Figure 1e), and such a merit is inherited by the GP/CNT composite, making it interesting for flexible/wearable electronics. Additionally, direct attachment of CNTs to the carbon surface potentially facilitates fast electron transport and thus good supercapacitor performance [9] in terms of high power capability.

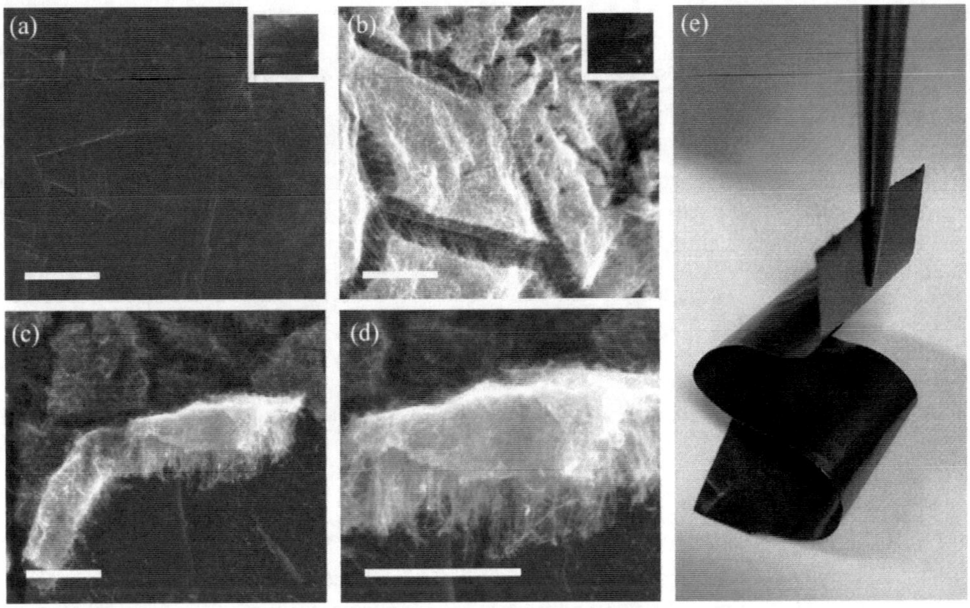

Figure 1. SEM images of (a) GP and (b-d) GP/CNT, and (e) demonstration of GP flexibility. The insets in (a-b) are the optical pictures of corresponding material. Scale bar in (a-d) is 1 μm.

Raman spectra were recorded in order to confirm the successful growth of CNTs and examine the CNT quality. Figure 2a shows the spectra of GP substrate and GP/CNT composite. GP spectrum is characterized by typical graphite features, with D, G, 2D bands at approx. 1330, 1580 and 2750 cm^{-1}, respectively. For GP/CNT, its G band clearly splits into G^+ and G^-. The doublet G band peak is attributed to a slight carbon sheet curvature and is considered as signature of CNTs.

The G band is derived from the carbon lattice stretching of the sp^2 bonded graphene plane, whereas the D band originates from the defect structure in the honeycomb graphene-like lattice [10]. The intensity ratio of G to D bands (I_G/I_D) is correlated with the amount of defects in graphitic materials, and can be used to reflect the quality of CNTs [11]. A mapping image of I_G/I_D within a selected ca. 20 × 20 μm^2 area is shown in Figure 2b. Representative spectra at the three specified spots are displayed in Figure 2c. The D, G^-, G^+ and G' band positions are identical for the three selected spots, with I_G/I_D ratio ranging from the lowest (5.7) to highest (7.7). Considering the non-continuous surface morphology (meaning that exposed graphite between cracks contributes to the G band signals), the Raman mapping data indicates a rather reasonable homogeneity of CNT quality.

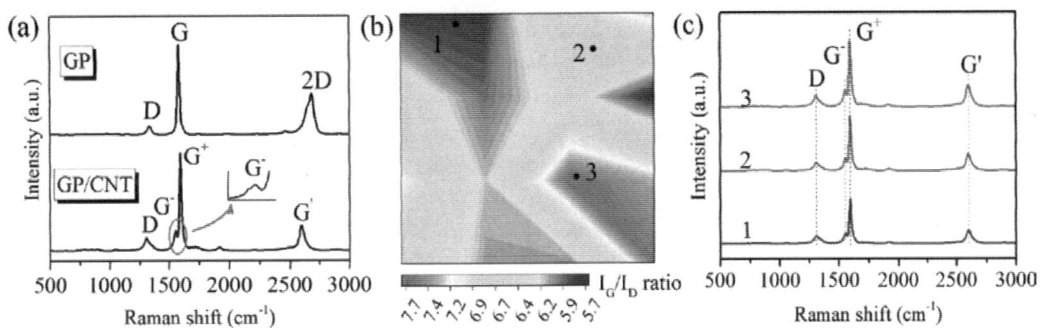

Figure 2. (a) Raman spectra of GP and GP/CNT, (b) Raman mapping of an approx. 20 × 20 μm^2 area on GP/CNT, and (c) comparison of Raman spectra at the three specified spots in (b).

IOP Conf. Series: Journal of Physics: Conf. Series **922** (2017) 012014 doi:10.1088/1742-6596/922/1/012014

3.2. Device performance

As previously mentioned, the fabricated GP/CNTs were used as electrodes in a supercapacitor for performance assessment. The CV curves are shown in Figure 3a-g. At lower scan rates, the device displays rectangular CV characteristics, which are indicative of double layer capacitive mechanism (commonly observed for carbon-based materials). With an increasing scan rate, the response current increases, while the CV curve correspondingly deviates from rectangular to quasi-rectangular shape. The current response remains linear to scan rates up to 1000 mV s^{-1} (Figure 3h), and the responsive CV curve is still symmetric up to 100000 mV s^{-1} (i.e. 100 V s^{-1}), indicating that the GP/CNT material has a good rate capability.

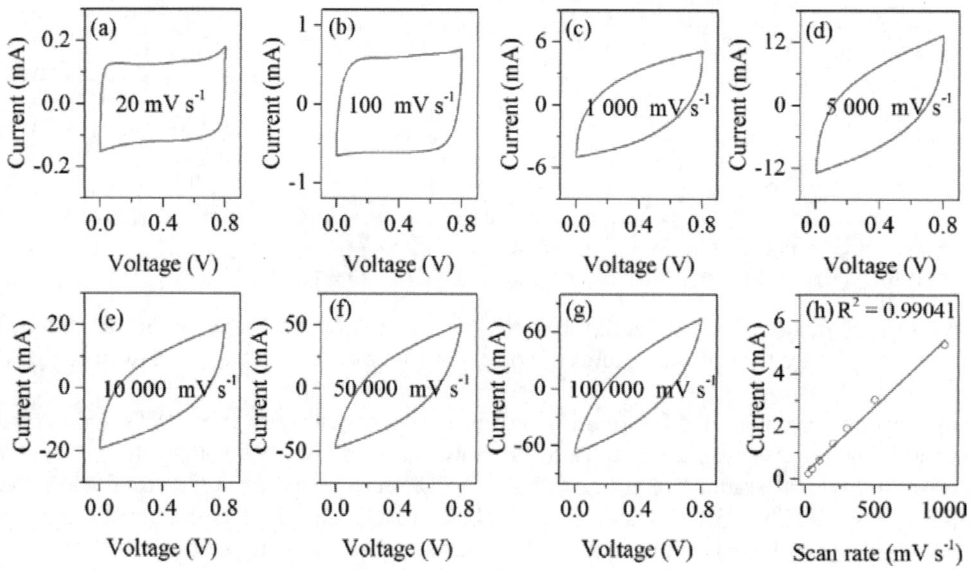

Figure 3. Cyclic voltammetry (CVs) of GP/CNT at specified scan rates (a-g), and (h) evolution of the cathodic current versus scan rate till 1000 mV s^{-1}.

GCD curves at different current densities are displayed in Figure 4a-b. In agreement with CV measurements, GP/CNT exhibits typical double layer GCD features of triangular curves. The highest areal capacitance of 4.4 mF cm^{-2} is achieved at the lowest current of 0.1 mA cm^{-2}. With the current density ramping up, the capacitance decreases to 0.8 mF cm^{-2} at a rather high current density of 15 mA cm^{-2} (Figure 4c). Accordingly, the maximum (0.38 μWh cm^{-2}) and minimum (0.04 μWh cm^{-2}) energy, and the minimum (29 μW cm^{-2}) and maximum (1790 μW cm^{-2}) power density are achieved at 0.1 and 15 mA cm^{-2}, respectively. It is worth noting that pure GP has much lower capacitance than GP/CNT as compared in Figure 4c, indicating that the energy of GP/CNT composite material mostly originates from the CVD grown CNTs.

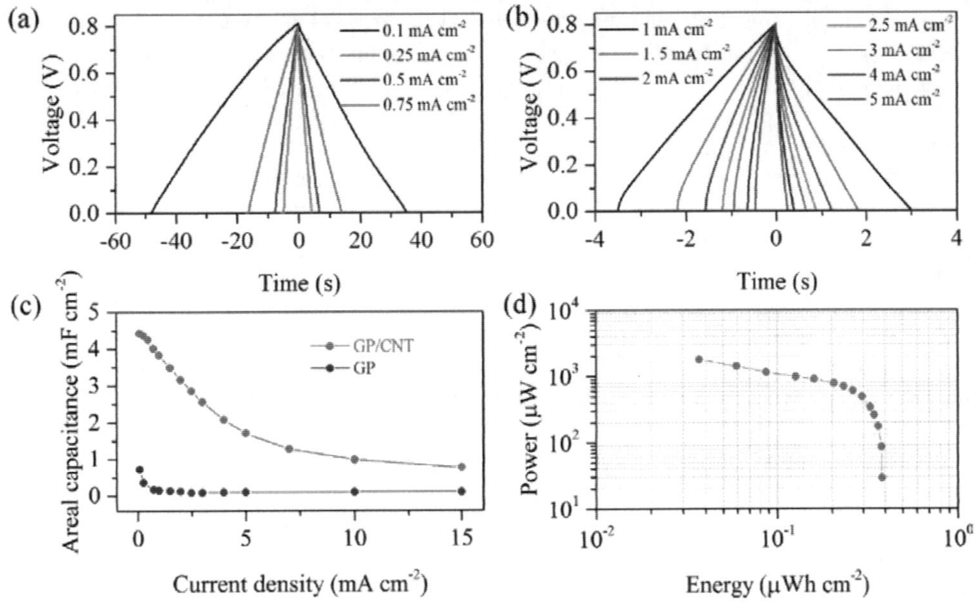

Figure 4. (a-b) Galvanostatic charge/discharge (GCD) curves of GP/CNT at different current densities; (c) The calculated areal capacitance of GP and GP/CNT at different current densities; (d) Ragone plot of GP/CNT.

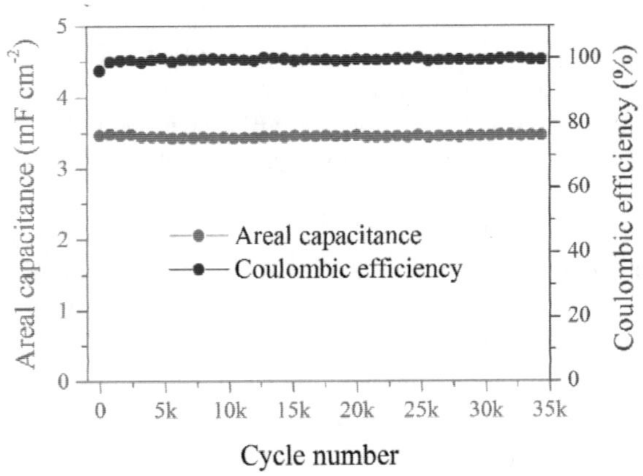

Figure 5. Cyclic charge/discharge (CCD) measurement at 2 mA cm^{-2}.

The GP/CNT-based supercapacitor also shows very high stability. The charge/discharge measurements repeated up to 35000 cycles with no observed decrease in capacitance. The coulombic efficiency (the ratio of discharge capacity to charge capacity) is close to 100% throughout the whole process.

4. Conclusions and future work

A freestanding flexible graphite paper composited with carbon nanotubes (GP/CNT) was successfully fabricated with the assistance of CVD-based CNT growth. The obtained GP/CNT material shows promising electrochemical performance as a supercapacitor electrode. The material flexibility makes the GP/CNT interesting as an energy storage component in flexible and wearable electronics as well as powering various microsystems. Future work can be directed into the following aspects: (1) Energy enhancement by increasing the CNT content or by compositing with pseudocapacitive materials; (2)

Rate capability improvement by removing the Fe/Al_2O_3 layer; (3) Fabrication of all-solid-state flexible supercapacitors; (4) Integration of supercapacitor components with other microsystem technologies such as sensors, MEMS harvesters and digital electronics.

Acknowledgements
This work was funded by the European Union's Horizon 2020 research and innovation program smart-MEMPHIS project (grant number 644378), and the Vinnova UDI project Energy Harvesting Toolkit.

References

[1] D. Heussner, Wearable Technologies Present Packaging Challenges, Electronic Design, (2014).
[2] X. Tao, Wearable electronics and photonics, Elsevier, 2005.
[3] Y. Zi, H. Guo, J. Wang, Z. Wen, S. Li, C. Hu, Z.L. Wang, An Inductor-Free Auto-Power-Management Design Built-in Triboelectric Nanogenerators, Nano Energy, 31 (2017) 302-310.
[4] M. Magno, T. Polonelli, F. Casamassima, A. Gomez, E. Farella, L. Benini, Energy-Efficient Context Aware Power Management with Asynchronous Protocol for Body Sensor Network, Mobile Networks and Applications, (2016) 1-11.
[5] Z. Fan, J. Yan, T. Wei, L. Zhi, G. Ning, T. Li, F. Wei, Asymmetric Supercapacitors Based on Graphene/MnO_2 and Activated Carbon Nanofiber Electrodes with High Power and Energy Density, Advanced Functional Materials, 21 (2011) 2366-2375.
[6] Q. Li, M. Haque, V. Kuzmenko, N. Ramani, P. Lundgren, A.D. Smith, P. Enoksson, Redox Enhanced Energy Storage in an Aqueous High-Voltage Electrochemical Capacitor with a Potassium Bromide Electrolyte, Journal of Power Sources, 348 (2017) 219-228.
[7] S. Jain, O. Yehia, L. Qiao, Flame Speed Enhancement of Solid Nitrocellulose Monopropellant Coupled with Graphite at Microscales, Journal of Applied Physics, 119 (2016) 094904.
[8] C.L. Pint, N.T. Alvarez, R.H. Hauge, Odako Growth of Dense Arrays of Single-Walled Carbon Nanotubes Attached to Carbon Surfaces, Nano Research, 2 (2009) 526-534.
[9] Y. Zhu, L. Li, C. Zhang, G. Casillas, Z. Sun, Z. Yan, G. Ruan, Z. Peng, A.-R.O. Raji, C. Kittrell, A Seamless Three-Dimensional Carbon Nanotube Graphene Hybrid Material, Nature Communications, 3 (2012) 1225.
[10] N.G. Yasri, A.K. Sundramoorthy, W.-J. Chang, S. Gunasekaran, Highly Selective Mercury Detection at Partially Oxidized Graphene/Poly(3,4-Ethylenedioxythiophene):Poly(Styrenesulfonate) Nanocomposite Film-Modified Electrode, Frontiers in Materials, 1 (2014).
[11] I.V. Anoshkin, I.I. Nefedova, D.V. Lioubtchenko, I.S. Nefedov, A.V. Räisänen, Single Walled Carbon Nanotube Quantification Method Employing the Raman Signal Intensity, Carbon, 116 (2017) 547-552.

Modular packaging concept for MEMS and MOEMS

Vanessa Stenchly[1], Wolfgang Reinert[1], Hans-Joachim Quenzer[1]
[1] Fraunhofer Institute for Silicon Technology, Itzehoe, Germany

vanessa.stenchly@isit.fraunhofer.de

Abstract. Wherever technical systems detect objects in their environment or interact with people, optical devices may play an important role. Light can be relatively easily produced and spatially and temporally modulated. Laser can project sharp images over long distances or cut materials in short distances. Depending on the wavelength an invisible scanning in near infrared for gesture recognition is possible as well as a projection of brilliant colour images. For several years, the Fraunhofer ISIT develops Opto-Packaging processes based on the viscous reshaping of glass wafers: First, hermetically sealed laser micro-mirror scanners WLP with inclined windows deflect in the central light reflex of the window out of the image area. Second, housing with lateral light exit permits hermetic sealing of edge-emitting lasers for highest reliability and durability. Such systems are currently experiencing an extremely high interest of the industry in all segments, from consumer to automotive through to materials processing. Our modular Opto-Packaging platform enables fast product developments. Housing for opto mechanical MEMS devices are equipped with inclined windows to minimize distortion, stray light and reflection losses. The hot viscous glass forming technology is also applied to functionalized substrate wafers which possess areas with high heat dissipation in addition to thermally insulating areas. Electrical contacts may be realized with metal filled vias or TGV (Through Glass Vias). The modular system reduces the development times for new, miniaturized optical systems so that manufacturers can focus on the essentials in their development, namely their product functionalities.

1. Motivation and concept

Miniaturized components must be generally protected against external influences, but not always, a polymer encapsulation is possible - mechanically moving elements or optical components must remain free, so that a cavity is required. The sticking of a package is not acceptable in many applications, because water vapor can pass the adhesive-seal and condense in the cavity which usually leads to a failure of the component [1]. Then the hermetic sealing of the entire assembly by welding or soldering with metallic-solder or glass frit is the only option. A hermetic package is inevitable, if special demands are made on the atmosphere in the assembly, for example, a defined vacuum, an inert gas filling or the complete absence of hydrocarbons [2]. The wafer-level packaging (WLP) has now become a standard technology in the field of micro-mechanical sensors; the packaging of optical components and systems on wafer level is still relatively rare. This is mainly due to the so far limited opportunities to make small optical housing by functional aspects or to meet all the requirements of optics, electronics and thermal management. This point is where the modular packaging system of Fraunhofer ISIT becomes important. The aim is to develop a "kit" for the production of hermetically sealed, application specific wafer level (WL) housing for the construction of optical and mechanical components and systems. The production of the glass cover is hereby based on high temperature

IOP Conf. Series: Journal of Physics: Conf. Series **922** (2017) 012015 doi:10.1088/1742-6596/922/1/012015

viscous glass micromachining with glass and silicon wafers [3], the micromechanical glass processing and a series of sequential bonding steps of glass and silicon wafers. This technology tool set enables the production of differently oriented, integrated windows on full 200 mm (8") wafers. However, the glass flow technology also enables the production of substrates with integrated electrical feedthroughs (Through-Glass Via, TGV) of highly conductive silicon (TGV A) or the introduction of heat sinks. Figure 1 illustrates the different possible combinations of different cap wafers and TGV Substrates and their applications.

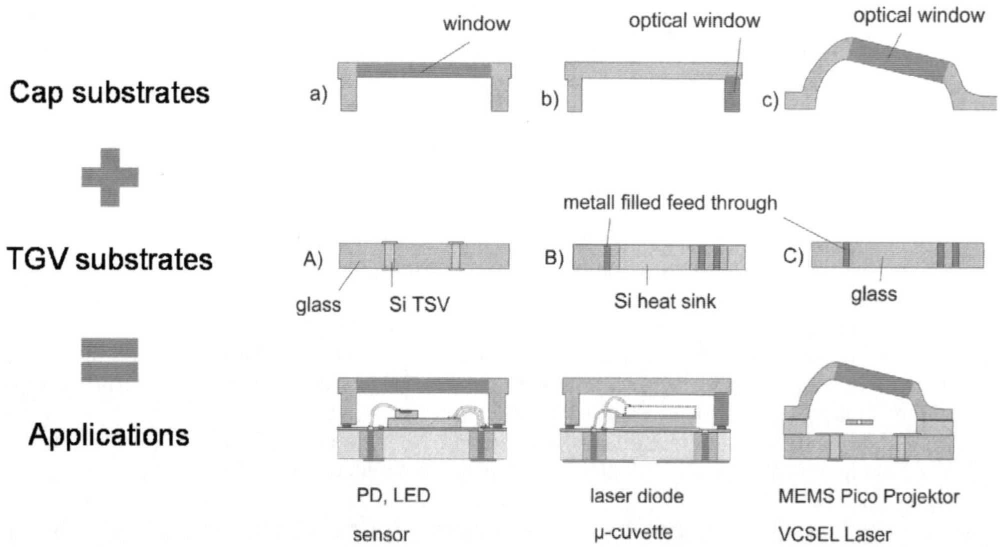

Figure 1: Schematic of different optical housing constructions supported by the modular packaging system.

2. Technology of different cap wafers

A new process has been developed for the fabrication of complete 8" cap wafer with a plurality of inclined optical surfaces for hermetic sealing of micro mirrors with the benefit of the elimination of disturbing reflections [4]. Figure 2 shows a schematic of the fabrication process flow in a cross sectional view. At first, deep cavities are etched into two silicon wafers with a DRIE process step (a). One of these two structured silicon wafers is anodically bonded on a borosilicate glass wafer (b). For this step it is necessary that all wafers are very clean and particle free. After anodic bonding a grinding step follows to define a pattern of silicon islands. The grinding process introduces residual stress in the silicon; a short stress relief step eliminates the surface stress in the silicon elements. Otherwise the stress in the silicon islands causes a warpage in the optical windows areas. This glass wafer with silicon structures on one side is finally bonded on a third silicon wafer, which provides pre-defined cavities wherein the silicon islands are inserted (c). The third silicon wafer is also structured by a DRIE etch process step. During the anodic bonding a nitrogen atmosphere with pre-adjusted pressure of about 950 mbar is enclosed in the cavities. After the anodic bonding the second silicon island wafer is bonded on the wafer stack (d). Then a last grinding step follows to define the islands on the outside of the glass wafer (e). That composite wafer is then annealed above the softening point of the glass material (f). By the thermally induced pressure difference between the cavities and the oven atmosphere above 680°C the silicon glass stacks inside the cavities are extruded within a defined time period. A non-centric arrangement of the silicon islands leads to an inclination of the silicon-glass stacks. Finally, the silicon is selectively removed in a KOH wet etch step (g). The final bonding of the processed glass wafer with the micro mirror wafer is done by the seal glass bonding. In Figure 3 a finished glass wafer with inclined windows is shown.

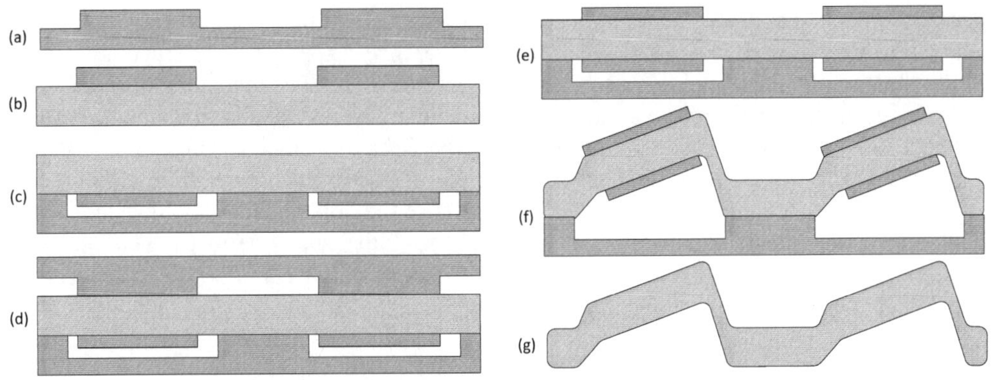

Figure 2: Schematic of the process flow for the glass wafer with inclined windows.

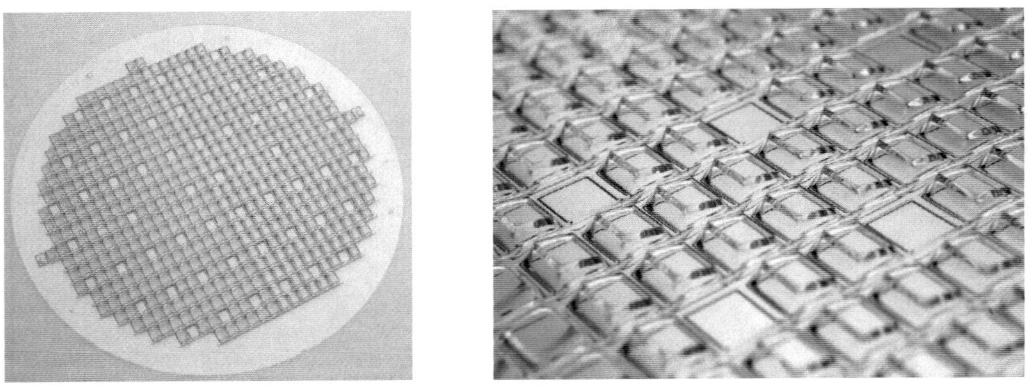

Figure 3: Complete 8” glass wafer with inclined windows and a detail view on a selected area.

Today laser diodes are packaged in TO housings that are not SMD compatible. Mobile applications require a much more economical and small factor package solution for light sources. We present a solution that realizes flat glass packages with vertical emission window for laser diodes on 8” silicon substrates. The new process uses a high temperature glass forming process with a combination of two different glasses each having different melting points. Figure 4 shows a schematic of the fabrication process flow for the optical cap with vertical emission window. At first, cavities are etched into a silicon wafer with a DRIE process step (a). After anodic bonding of this structured silicon wafer with a borosilicate glass wafer (b), a mechanical dicing step follows from the silicon side along the DRIE etched silicon structures to define the pinboard (c) wherein the high melting glass window can be inserted later (d). The height of the silicon islands is predetermined by the wafer thickness of the starting material and also defines the cavity height. After a cleaning step to remove silicon residuals the high melting glass stripes can be applied (e). Because the stripes were previously separated from a wafer by mechanical dicing, they have to be annealed in a separate step to take out the moisture that accumulates in the diced edges. When the stripes will be inserted in the pinboard they have to be turned upright 90°, so the polished and smooth surface is arranged perpendicular to the substrate surface. After a further silicon wafer is applied, the complete stack is rotated. In the next step the glass-edge of the peripheral area is to be depressed in vacuum, so the glass is hermetically sealed with the silicon. Thereby the glass seals with the silicon uniformly and without air pockets, a ROBAX®-ring coated with boron nitride is placed on the stack to work as a downholder (f). Here, it is important that the annealing is performed under the transformation temperature of the high melting glass (g). In a second step, the stack is annealed under atmospheric pressure at the same temperature until the glass is completely flown or pressed down by the pressure difference in-between the silicon structures and fuses with the high melting point glass window sections (h). The annealing duration is highly dependent on the design and the aspect ratio of the silicon structures. Finally, the silicon is selectively removed in a KOH wet etch step (i). After releasing the complete silicon from the cap wafer a two sided grinding step follows. The front side has to be planarized to enable the following processes. A grinding step on the lower side must be carried out, because the borosilicate glass does not completely wet the high-melting glass. This resulting gap prevents subsequent hermetic sealing. After this planarization an anti-reflective coating is applied. In the next step a plating base is deposited with a shadow mask on the bond frames of the cap wafer (Fig. 5).

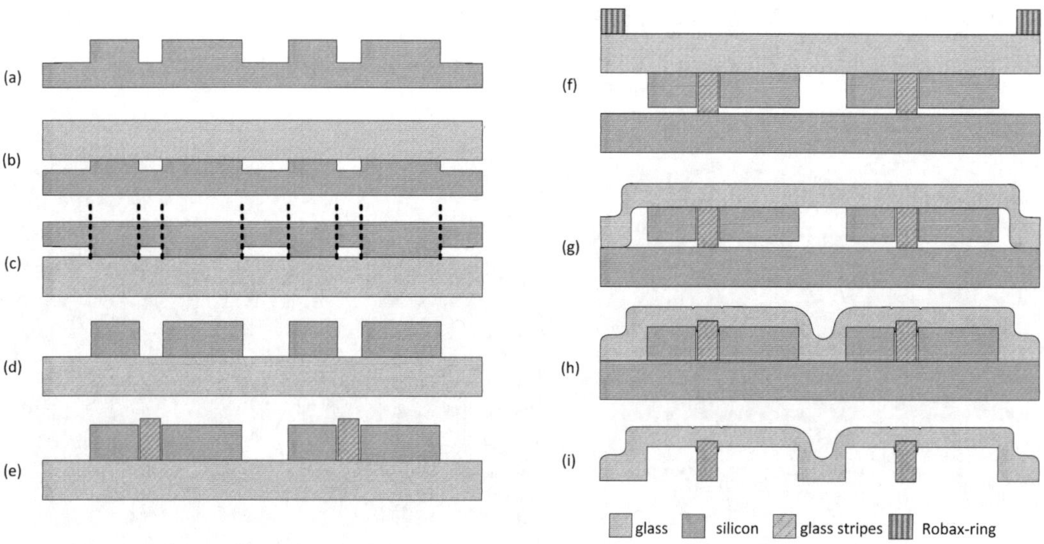

Figure 4: Schematic of the process flow for the glass wafer with vertical windows.

Figure 5: Left: Silicon wafer with inserted glass stripes after second annealing. Right: Detail view of a glass cap wafer with vertical emission windows.

3. Technology of the Through Glass Vias Wafer

To keep the entire component as small as possible, electrical contacts may be realized by Through-Glass-Vias. The schematic in Figure 6 (left) shows our process flow which limits the waferbow to less than 100 µm [5]. At first, deep cavities with a high aspect ratio are etched into a silicon wafer with a DRIE process step (a). After anodic bonding with a glass wafer (b) an annealing step follows where the glass is completely pressed in the small structures (c). To eliminate the unevenness of the wafer surface, the wafer is grinded until the silicon contacts are opened (d). Then the backside of the wafer is also grinded to remove the silicon. Finally the TGV is polished to remove the surface roughness. In Figure 6 (right) a part of a glass substrate wafer with silicon vias is shown.

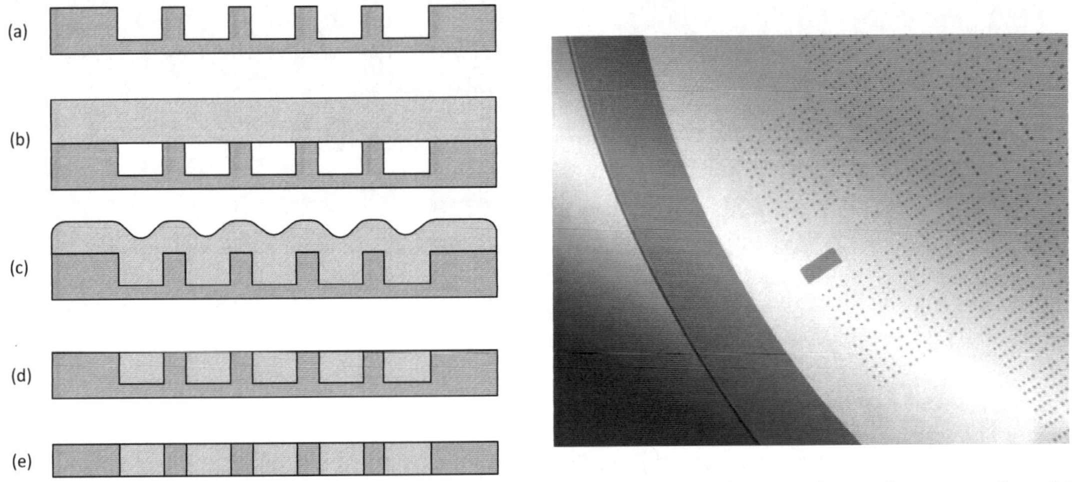

Figure 6: Schematic of the process flow for TGV-wafer (left) and view on glass substrate wafer with silicon vias (right).

4. Examples of packaged laser diodes and MEMS mirrors

The cap wafers with inclined windows are typically bonded to micro mirror wafers by using glass frit sealing (Fig. 7, left). Figure 7 (right) shows a cross section of an X-Ray photograph of a fully caped mirror device. The inclined optical window and the surrounding socket are clearly visible. The integrated cavity ensures the free movement of the enclosed micro mirror and the bonding of an additional spacer wafer could be avoided.
The first processed glass cap wafers with a vertical emission window for laser diodes (side looking) and the glass cap wafers with inclined windows for micro mirrors were built up without the TGV-wafer. The packaged laser diode demonstrators were assembled on silicon based wafer substrates. AuSn metallized solder pads and metal seal frames enable an eutectic submount- and laser diode soldering. An active solder was used for hermetic bonding of the single glass cap with the silicon substrate (Fig.8).

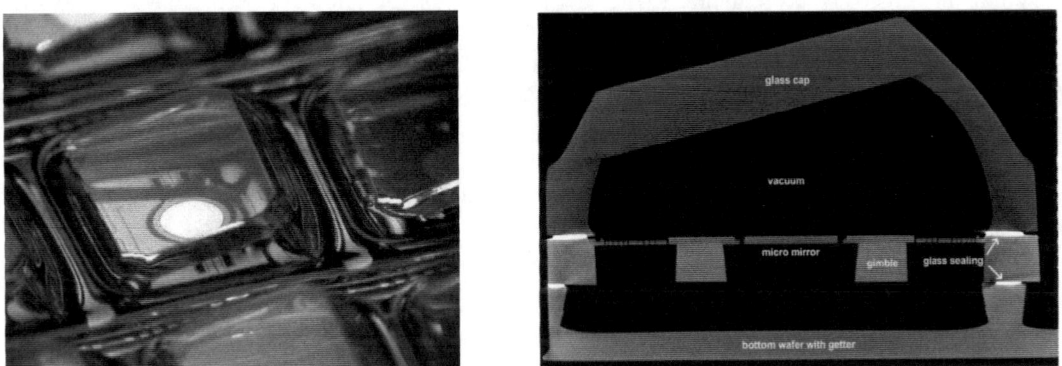

Figure 7: Close up of a wafer level vacuum packaged mirror with an inclined window with a size of 3.3 x 4.5mm (left) and a X-Ray cross section through a full-caped-mirror-device (right).

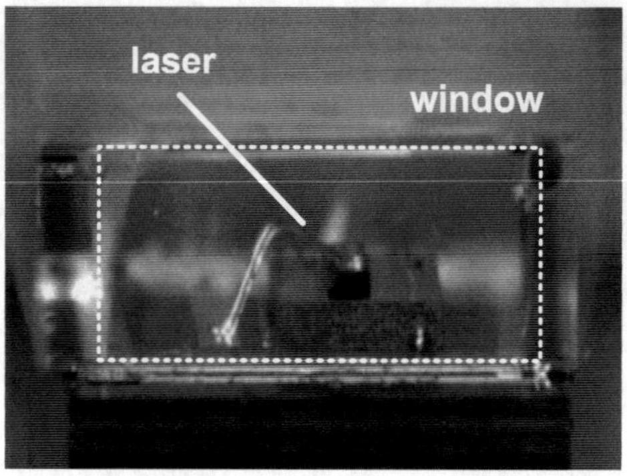

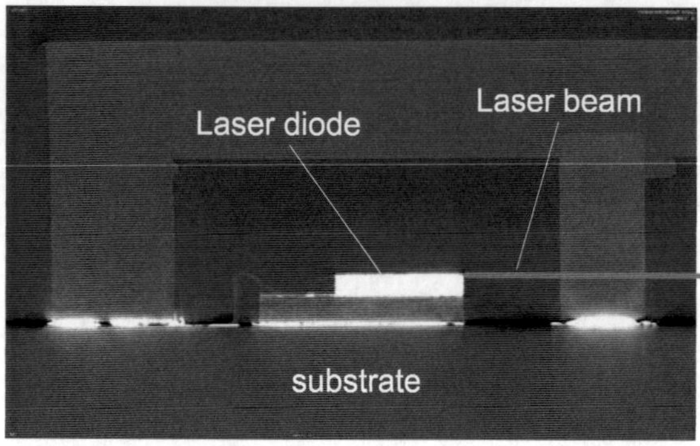

Figure 8: Detail view of the glass cavities within an 8" wafer with vertical emission windows (left) and a X-Ray cross section through a housed laser diode (right).

5. Conclusion

The first processed glass cap wafers for laser diodes and micro mirrors demonstrate the capabilities of the new developed glass forming technologies. Recently the first laser diodes and micro mirrors have been successfully packaged with the developed glass caps. The actual work is focused on the integration of the TGV wafer and the expansion to further glass cap processes.

Acknowledgements

Parts of the work have been performed in the project PICOLO, funded by Federal Ministry of Education and Research (BMBF 13N12426).

Patents pending

The presented technologies are patented under US 7.416.951 B2 and US 14.361.392.

References

[1] V. Lindroos, M. Tilli, A. Lehto, T. Motooka, Silcon Based MEMS Materials & Technologies, Part V Encapsulation of MEMS components, Elsevier, 2010.

[2] M.Oldsen, U.Hofmann, H.J. Quenzer, B. Wagner, A Novel Fabrication Technology for Waferlevel Vacuum packaged Microscanning Mirrors, Proceedings of the 9th Electronics Packaging Technology Conference, Singapore, 2007.

[3] P. Merz, H. -J. Quenzer, H. Bernt, B. Wanger, M. Zoberbier, A novel micromachining technology for structuring borosilicate glass substrates, TRANSDUCERS, Solid-State Sensors, Actuators and Microsystems, 12th International Conference, vol.1, 2003, pp 258 – 261.

[4] V. Stenchly, H. -J. Quenzer, U. Hofmann, J. Janes, B. Jensen, W. Benecke, New fabrication method of glass packages with inclined optical windows for micro mirrors on wafer level, Proceedings of the SPIE, Vol. 8613, 861319, 2013.

[5] V. Stenchly, W. Reinert and H.-J. Quenzer, Hermetic packaging concept for laser diodes on wafer level, Proceedings of the Electronic System-Integration Technology Conference (ESTC), 2016 6th.

Dextran as a fast resorbable and mechanically stiff coating for flexible neural probes

D Kil, L Brancato and R Puers

KU Leuven dept. ESAT-MICAS, Kasteelpark Arenberg 10, 3001 Leuven, Belgium

Abstract. In this paper we report on the use of dextran as a temporary, fast dissolving stiff coating for flexible neural probes. Although polymer-based neural implants offer several advantages, compared to their rigid silicon counterparts, they pose significant challenges during implantation. Due to their extreme flexibility, they have the tendency to buckle under the axial load applied during insertion. The structural stiffness of the implants can be temporarily increased by applying a bioresorbable dextran coating which eases the penetration of neural tissue. For this application three types of dextran with different molecular weights are analysed. The dissolution rate of the coatings is reported as well as the increased bending stiffness resulting from the dextran coating of Parylene C neural probes. Based on these findings the dissolution rate can be linked to parameters such as molecular weight, coating thickness and the surface area exposed to the dissolution medium. The mechanical characterization yields information on how the structural stiffness of neural probes can be tuned by varying the dextran's molecular weight and coating thickness.

1. Introduction

It is generally known that in the field of neural engineering there is a trend towards the use of flexible polymers as the structural material for neural probes [1]. When neural implants are made out of hard, stiff materials such as silicon [2] these have the tendency to move around under influence of, among others, blood being pumped through the arteries in the brain. These micromotions of the electrode relative to the brain result in a chronic inflammation of the tissue surrounding the implant, which in turn leads to glial scar formation [3]. The dense scar tissue formed around the electrodes effectively isolates them from any surrounding active neurons, interfering with recording or stimulation and rendering the implant useless after a relatively short period of time [4]. This problem can be partially solved by using polymeric materials such as parylene [5] and polyimide [6] which reduce the mechanical mismatch with the surrounding brain tissue and deform as the tissue moves. The downside of this approach is that these thin polymer implants are too flexible to be able to penetrate the neural tissue without buckling.

A frequently used solution is the use of a stiff backbone structure. Examples include stiff backbone layers [7], insertion shuttles [8] and biodegradable coatings [9]. Although these options temporarily increase the structural stiffness of the implant they do come with drawbacks. The stiff backbone layer is intended to remain permanently attached to the implant, it significantly reduces its innate flexibility. Insertion shuttles on the other hand temporarily increase the footprint of the implant, resulting in unnecessary damage to the neural tissue during implantation. The least invasive solution is to temporarily package the implant using a bioresorbable coating such as silk [10] or PEG [11] which stiffens it enough to be implanted. Afterwards the coating dissolves, allowing the implant to regain its flexibility. A downside to this technique, which should be kept in mind, is that in most cases the coating dissolves so rapidly that there is only one opportunity for implantation.

In this paper, we research the use of dextran for this purpose. Apart from its use in medicine as an antithrombic or volume expander for hypovolaemia [12] it also has excellent functionality as a temporary coating for neural probes. Dextran not only limits non-specific cell adhesion [13] but as it is a complex branched glucan its dissolution rate and mechanical stiffness can be tuned by varying the chain length of the molecule.

Compared to the materials mentioned above, comparable data on the use of dextran as a coating material is not readily available in literature. This work tries to fill that gap and presents an extensive characterization of dextran as a suitable material for this purpose. Ultimately the goal of this paper is to act as a reference guide containing information on how the parameters mentioned above relate to the mechanical stiffness and dissolution rate, allowing the reader to tune his dextran coating to application specific needs, albeit dissolution time or structural stiffness.

2. Experimental details

2.1 Density measurement

Dextran slabs were prepared by dissolving the powder in de-ionized water, creating a highly viscous solution (50 wt%). The mixture was stirred under vacuum to remove any airbubbles. After degassing, the solution was poured into a teflon mould an allowed to dry at low temperature (50°C). When all the water evaporated the dextran slabs where demoulded and weighed.

The volume of the dextran slabs was determined using a water displacement experiment. The samples were immersed in water in a graduated cylinder, and their volume was estimated measuring the increase in the water level. The average density is calculated and summarized in table 1. This experiment was repeated to determine the density of 3 types of dextran with varying molecular weights: 40 kDa, 100 kDa and 500 kDa.

Table 1. Average density of different types of dextran.

Dextran molecular weight (kDa)	40	100	500
Average density (g/mm^3)	2.5E(-3)	2.3E(-3)	1.4E(-3)

2.2 Dextran dissolution rate

Dextran slabs were prepared in a square stainless steel mould as described in section 2.1. After drying, the samples were weighed and submerged in deionized water for a known amount of time. At regular time intervals the samples were removed from the container, allowed to dry, and weighed. This provided an indication on the quantity of dextran dissolved over time. The dextran slabs were submerged while still in the mould in order to keep the surface area exposed to the dissolution medium constant in time, which results in a more accurate representation of the dissolution rate. This experiment was also repeated for the different types of dextran with varying molecular weights.

2.3 Micromechanical testing

In order to quantify the influence of the dextran coating on the mechanical properties of the implants, a set of dummy Parylene C shanks was fabricated and coated using a dipcoating process. The thickness of the dextran coating was measured using a digital micrometer.

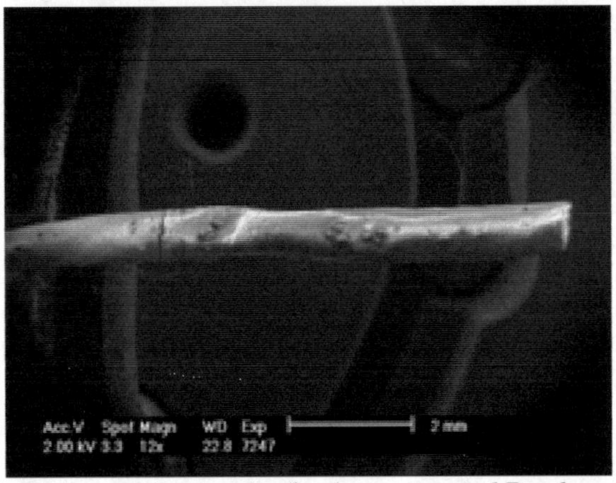

Figure 1. Electron micrograph of a dextran coated Parylene C shank.

The samples were glued to an aluminum oxide substrate, leaving a known part the shank sticking out over the edge. A Femtotools FTA-M02 micromechanical testing system was used to displace the freestanding part of shank. Force measurements were taken during this displacement and used to determine the bending stiffness of the samples. The out-of-plane bending mimics the forces that the shank undergoes during implantation. The Young's modulus E of the samples approximated using the formula for bending of a beam under a point load:

$$E = \frac{k \,.L^3}{3 \,.I} \tag{1}$$

With I the area moment of inertia:

$$I = \frac{w \,.h^3}{12} \tag{2}$$

In these formulas, L represents the distance between the hinging point and the contact point of the indenter tip, w is the measured width of the sample and h is the sample thickness.

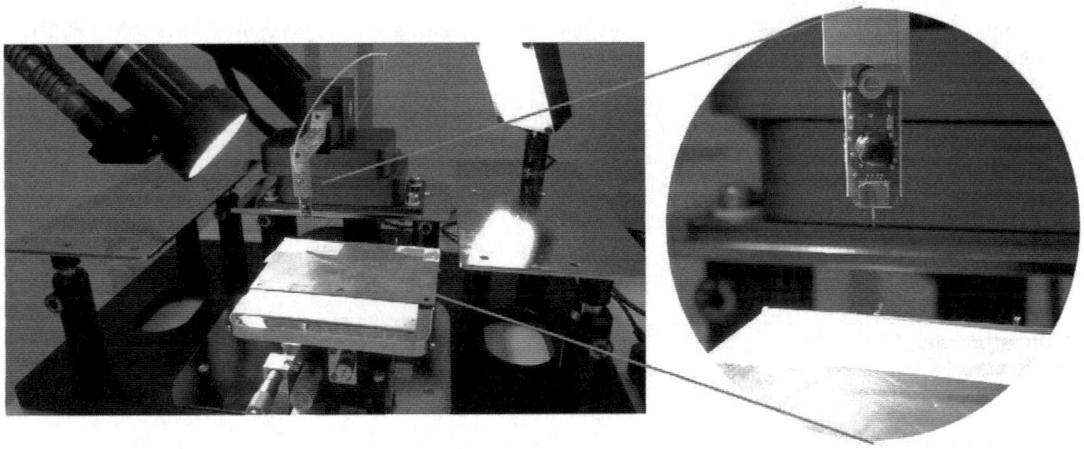

Figure 2. Test setup for measuring bending stiffness using the Femtotools FTA-M02 micromechanical testing system. Close-up shows the vertical indenter probe with integrated force sensor.

3. Results and Discussion

3.1 Dextran dissolution rate

The results of the dissolution experiment are depicted in fig. 3. It is clearly visible that the higher the molecular weight of the dextran, the lower the weight loss over time is. This indicates a strong dependency between dissolution rate and molecular chain length which can be explained by the physical mechanism of sugar dissolution in water.

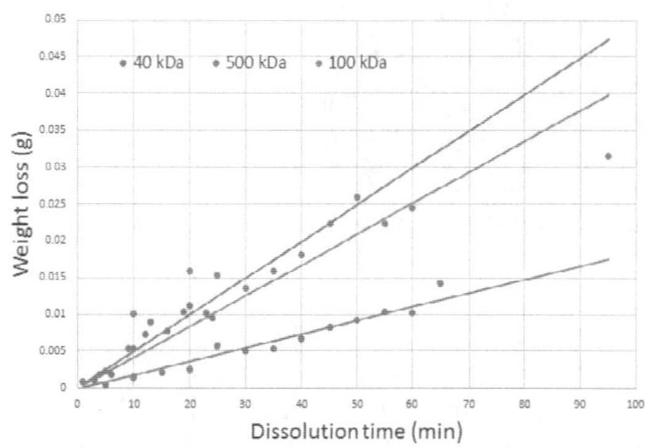

Figure 3. Dextran dissolution rate under influence of molecular weight.

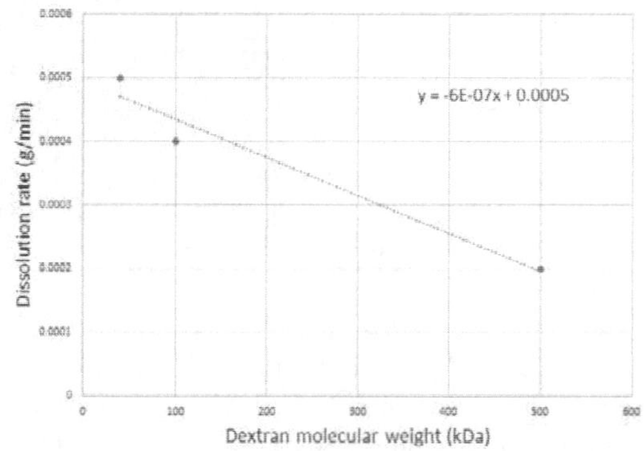

Figure 4. Dextran dissolution rate in function of molecular weight.

Sugar molecules contain a high number of polar oxygen-hydrogen bonds, which have a permanent dipole moment. In the solid crystal structure itself a number of these molecules are held together by weak intermolecular Vanderwaals-forces. When the sugar crystal comes into contact with water, the polar bonds of the water molecules form their own dipole-dipole bonds with the sugar molecules, which are stronger, causing them to separate and bond to the water molecules.

The longer the molecule chain is, the more difficult it is for the water molecules to form dipole-dipole interactions with the functional groups of the sugar. As the folded molecules are much bigger, the surface area of the molecule exposed to the solute is relatively small compared to its size. This effect increases exponentially with the molecular weight and thus the dissolution rate will drop decrease as the molecular weight of the sugar increases. This physical dissolution process offers us the opportunity to tune the dissolution rate of our resorbable coating by varying the molecular weight of the dextran as is shown in figure 4.

It is important to note that these 3 types of dextran can only be compared because the surface area exposed to the dissolution medium was kept constant during the whole experiment. If the surface area is increased, more dextran molecules would come into contact with the medium, increasing the rate at which the coating dissolves. This linear relationship is depicted in figure 5.

Table 2. Average dissolution rate of dextran with varying molecular weights.

Molecular weight (kDa)	40	100	500
Average dissolution rate (mg/min*mm^2)	0.02	0.016	0.008

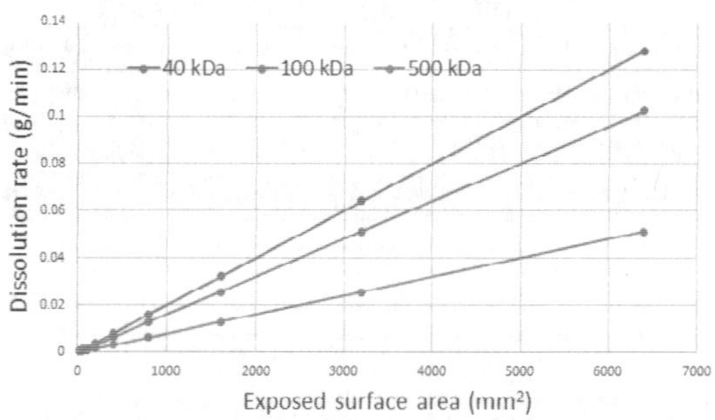

Figure 5. Dextran dissolution rate in function of the surface area exposed to the dissolution

As the dissolution mechanism is considered as a 'layer-by-layer' process the thickness of the coating has no influence on the dissolution rate. As we now have information on all the factors influencing the dissolution time of a dextran coating a formula is presented based on which the dissolution time of an arbitrary dextran coating can be calculated.

$$dissolution\ time = \frac{density * coating\ volume}{dissolution\ rate * surface\ area} \tag{3}$$

The density of the coating is dependent on the molecular weight of the used dextran and can be interpolated from the data presented in table 1. The volume of the coating can be measured and the dissolution rate can be deduced from figures 4 and 5, knowing the molecular weight of the used dextran and the surface area exposed to the dissolution medium.

3.2 Micromechanical testing

Measurements were taken of samples coated with the different types of dextran as described in section 2.3 and are summarized in table 3. The thickness measurements are also reported on the graph. The average width of the coating layer was 641 μm. Micromechanical testing revealed an average bending stiffness of 0.029 $N.m^{-1}$ for the uncoated implants. Coating increased this average bending stiffness 361 fold to 10.5 $N.m^{-1}$. The Young's modulus of coated sample was measured to be only 0.6 GPa, proving that the enhanced structural stiffness can be mainly attributed to the thickness of the coating and not to the molecular weight of the used dextran.

Table 3. Results of mechanical measurements

	Mean	**Standard deviation**
Coating thickness (μm)	163	21
Bending stiffness(N.m-1) uncoated	0.029	0.005
Bending stiffness(N.m-1) coated	10.5	5.5
Young's Modulus Parylene C (Gpa)	2.8	0.4
Young's Modulus dextran (Gpa)	0.6	0.1

4. Conclusion

By analysing the dissolution rate and mechanical properties of different types of dextran, its use a biocompatible, stiff and fast resorbable coating for neural implants was proven. A dissolution experiment was performed to determine the dissolution rate of the coatings obtained from dextran's with three different molecular weights. The results showed a linear dependency of the dissolution rate from the molecular weight. The dissolution rate is influenced from the surface area exposed to the dissolution medium. The enhanced structural stiffness resulting from dextran coating of neural probes was investigated using a Femtotools micromechanical testing system. Results showed a 361-fold increase in bending stiffness compared to an uncoated Parylene C probe. Finally, a formula is presented which relates all influencing parameters and allows to tune the dissolution time and mechanical properties of dextran coatings.

IOP Conf. Series: Journal of Physics: Conf. Series **922** (2017) 012016 doi:10.1088/1742-6596/922/1/012016

Acknowledgements
The research leading to these results has received funding from the European Research Council under the European Union's Seventh Framework Programme (FP7/2007-2013) / ERC grant agreement n° 340931, and the KU Leuven IDO program, project on electrical brain stimulation IDO/12/024.

References

[1] Weltman A, Yoo J and Meng E 2016 Flexible, penetrating brain brobes enabled by advances in polymer microfabrication *Micromachines* **7** 180
[2] Yoon T H, Hwang E J, Shin D Y, Park S I, Oh S J, Jung S C, Shin H C and Kim S J 2000 A micromachined silicon depth probe for multichannel neural recording *IEEE Trans. Biomed. Eng.* **47**, 1082–7
[3] Biran R, Martin D C and Tresco P A 2007 The brain tissue response to implanted silicon microelectrode arrays is increased when the device is tethered to the skull *J. Biomed. Mater. Res.* **82** 169–78
[4] Turner J N, Shain W, Szarowski D H, Andersen M, Martins S, Isaacson M and Craighead H 1999 Cerebral astrocyte response to micromachined silicon implants *Exp. Neurol.* **156** 33–49
[5] Takeuchi S, Suzuki T, Mabuchi K and Fujita H 2004 3D flexible multichannel neural probe array *J. Micromech. Microeng.* **14** 104–7
[6] Fomani A A and Mansour R R 2011 Fabrication and characterization of the flexible neural microprobes with improved structural design *Sensors Actuators A* **168** 233–41
[7] Lee K, He J, Singh A, Massia S, Ehteshami G, Kim B and Raupp G 2004 Polyimide-based intracortical neural implant with improved structural stiffness *J. Micromech. Microeng.* **14** 32-37
[8] Kozai T D Y and Kipke D R 2009 Insertion shuttle with carboxyl terminated self-assembled monolayer coatings for implanting flexible polymer neural probes in the brain *J. Neurosci. Methods* **184** 199-205
[9] Li W, Rodger D C, Pinto A, Meng E, Weiland J D, Humayun M S and Tai Y C 2011 Parylene based integrated wireless single channel neurostimulator *Sensors Actuators A* **166** 193-200
[10] Tien L W, Wu F, Tang-Schomer M D, Yoon E, Omenetto F G and Kaplan D L 2013 Silk as a multifunctional biomaterial substrate for reduced glial scarring around brain-penetrating electrodes *Adv. Funct. Mater.* **23** 3185-93
[11] Chen C-H, Chuang S-C, Su H-C, Hsu W-L, Yew T-R, Chang Y-C, Yeh S-R and Yao D-J 2011 A three-dimensional flexible microprobe array for neural recording assembled through electrostatic actuation *Lab Chip* **11** 1647-55
[12] Jones C, Payne D, Hayes P, Naylor R, Bell P, Thompson M and Goodall, A 2008 The antithrobotic effect of dextran-40 is due to enhanced fibrinolysis in vivo *Journal of Vascular Surgery* **48** 715-22
[13] Massia S P, Stark J and Letbetter D S 2000 Surface-immobilized dextran limits cell adhesion and spreading *Biomaterials* **21** 2253-61

Pick and Place Assembly Technique for Fabrication of Groove Gapwaveguide Resonator

Sadia Farjana[1], A.U. Zaman[2], Peter Enoksson[1]

[1]Micro- and Nanosystems Group, Electronics Materials and Systems Laboratory, Department of Microtechnology and Nanoscience (MC2), Chalmers University of Technology, 412 96 Göteborg, Sweden.
[2]Department of Signals and Systems, Chalmers University of Technology, SE-412 96 Gothenburg,Sweden

farjana@chalmers.se

Abstract. Gap waveguide are an innovative technology for millimeter-wave RF applications. There are some factors to be considered while using this technology at millimeter wave frequencies or above, particularly for low cost RF applications. Surface mount technology and pick and place machine technique can demonstrate a way to solve this problem. In this paper a groove resonator is presented which is manufactured by using pick and place technique. Brass pins of diameter 1.6mm and height 2.3mm was mounted on a 5mm thick copper plate. Lead free solder paste was used for solder joint.

1. Introduction

Surface-mount technology (SMT) and surface mount device (SMD) is a hot topic in electronic society[1] . SMT is a method for producing electronic circuits in which the components are mounted or placed directly onto the surface of printed circuit boards (PCBs). Pick and place systems or pick and place machines are usually used in the manufacture of surface mount technology (SMT) electronics circuit boards. Using these pick and place systems or pick and place machines, it is possible to accurately place large numbers of small, or large components quickly and precisely onto circuit boards. Practically all of today's mass produced electronics hardware is manufactured using surface mount technology. The associated SMDs provide many advantages over their leaded predecessors in terms of manufacturability and often also performance.

The blend of millimeter-wave communications, arrays with a substantial number of antenna elements, and shorter coverage distances are the likely factors that will allow merging of different technologies with the aim to dramatically improve wireless access and throughput in the upcoming years [2-4]. Even though a lucrative license free spectrum is available at millimeter-wave frequency range, still there are plenty of technological factors and mechanical challenges in designing millimeter-wave RF front-ends. These factors are cost, smaller size requirement, increased system density, packaging and cross-talk suppression, and lower power loss dissipation [5].

Conventional rectangular waveguides, planar transmission lines such as coplanar waveguides, or microstrip lines are well characterized transmission medium which are used in a variety of complex RF component and circuit designs until today. However, there are some factors to be considered while using these conventional technologies at millimeter wave frequencies or above, particularly for low cost RF applications. Most of the millimeter-wave wireless communication systems need analog RF hardware such as waveguide based filters and high gain waveguide based antennas. These waveguide based RF components are usually not produced by pick and place assembly technique and need another more costly manufacturing technique such as metal molding or metal milling. Thus, the total production cost for most of the RF wireless systems cannot be reduced below a certain level around the millimeter-wave frequency range.

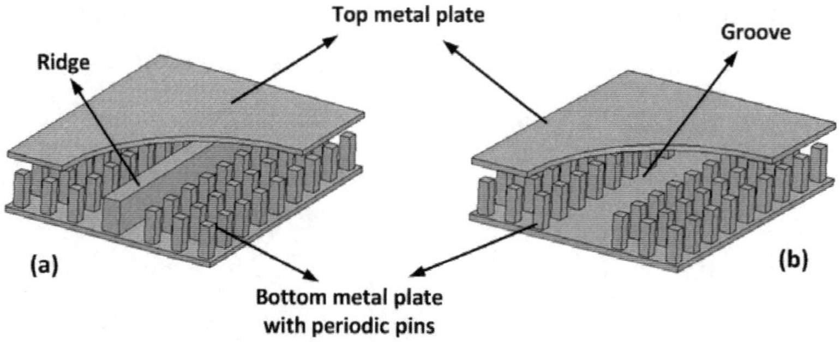

Figure 1. Basic Gap waveguide geometry: a) ridge gap waveguide; b) groove gap waveguide.

Recently evolved Gap waveguide are considered as an innovative technology for millimeter-wave RF applications [6,7,8]. As described in [8] and shown in Fig. 1, the basic gap waveguide structure consists of two metal parts; the bottom part consists of periodic metal pins or fingers and the top part is a smooth metal plate. Unlike the traditional rectangular waveguide, the top and bottom metal parts in any of these gap waveguide geometries do not need metal connections or electrical contacts. This opens up a completely new opportunity for designing and manufacturing RF components based on gap waveguide technology. It is expected that the bottom plate having a base metal and protruding fingers or ridge section can be produced very cheap by stamping on a soft thin metal sheet or by die-casting technique. These new manufacturing techniques will make the gap waveguide components relatively cheaper than the conventional rectangular waveguide based components. Nevertheless, the incompatibility of these manufacturing techniques with the pick and place assembly technique will still present a major bottleneck in integrating the digital electronics, RF electronics and the waveguide components in most of the millimeter wave wireless systems. Thus, the total production cost will not be reduced below a certain level. To overcome this issue, we present the concept of manufacturing a millimeter-wave RF waveguide component by using the pick and place assembly technique in this work. In the gap waveguide design, the size of the basic building blocks such as the metal pins, the metal ridge etc. falls in the range of typical size of the conventional SMDs. Pre-produced gap waveguide building blocks or basic components in large scale volume will make it possible to assemble these components by pick-and-place methodology. These pre-produced gap waveguide building blocks can be picked and placed on top of a carrier substrate and can be connected conductively to the conducting layer of the substrate. If several of these pre-produced components are picked and placed or put in a desired arrangement, a complete RF waveguide component can be formed. Thus making the manufacturing very adorable to different needs. The top metal plate for such gap waveguide component can be a simple metal plate or another substrate with ground plane. This new concept of pick and place type waveguides has the potential to revolutionize the millimeter wave RF industry completely and will pave the way for low cost millimeter wave RF products such as millimeter wave massive MIMO modules.

2. Groove gap waveguide resonator

To verify the performance of the low loss gap waveguide technology, it is important to characterize its loss in an accurate way. One way to do that is to measure the attenuation of a very long line (several wavelengths), which will be impractically long in order to measure the low losses accurately. As an alternative, losses can be better characterized in terms of Q-factors of resonators. One such groove gap resonator working at 30 GHz band is shown in figure 2(a) and 2(b).

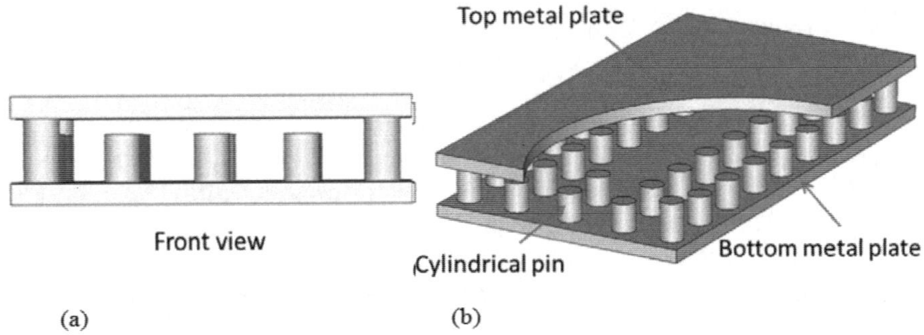

Figure 2. Schematic of a 30 GHz groove resonator. (a) Side view, (b) perspective view with part of the top plate cut away for clarify.

The dimensions of this groove gap resonator are shown in Table 1. The simulation result of the resonator is shown in Fig. 3. For a two-port resonator S21 is the transmission coefficient of the resonator (ratio of the voltage transmitted to the incident voltage). For Q factor evaluation, weak coupling has been considered in this case. The material used in simulation is copper. The simulated Q value at 27.75 GHz was found to be 1,800. In this paper we present a groove resonator that has been manufactured by using pick and place machine to realize this design. Pins have been mounted on the bottom plate by using the reflow soldering technique.

Table 1. Dimensional details of the groove gap resonator

Parameters (mm)	**Operating Frequency 30 GHz**
Pin Height	2.6
Pin Diameter	1.8
Gap between two pins	2.6
Resonator Length	19.2
Resonator width	2.6
Air Gap	.75

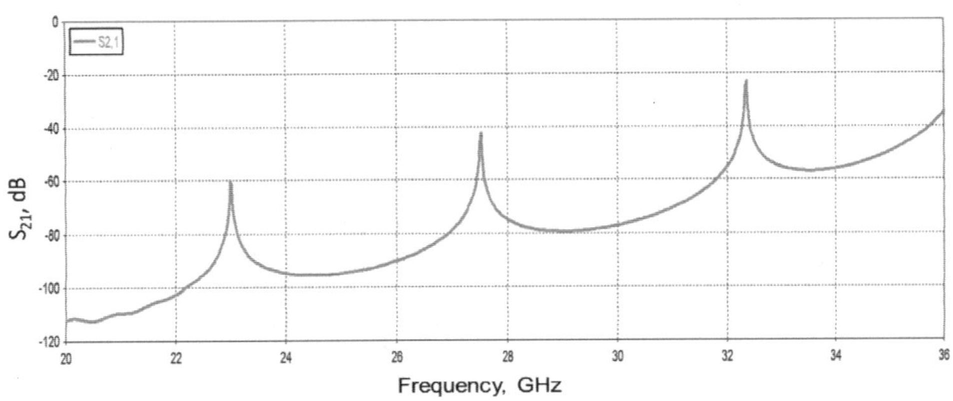

Figure 3. Simulation result of the 30 GHz groove resonator with several resonance peaks within the frequency band of interest.

3. Experimental details

A copper plate of 5 mm thickness was used as top and bottom plate and cylindrical brass pins of diameter 1.6 mm and height 2.3 mm were used for mounting. Before applying solder paste, flux was applied on the bottom plate. Lead free solder paste Sn96.5Ag3Cu was used as solder paste. The solder paste was applied by using a pump dispenser. Then a manual pick and place machine (for mass production automated machine will be used) was used to place the pins on the bottom plate. After placing the pins, the copper plate was placed in a reflow oven (PortoFlow S Solectro) for reflow soldering. During reflow soldering four different temperature zone i.e. preheat zone, soak zone, and reflow zone and cooling zone was set. The device was keep in the preheat zone for 200 s at 160 °C then kept in the soak zone for 240 s and then keep at reflow temperature 270 °C for 160 s and then in the cooling zone for 80 s. The reason for using four different temperature zone is described in section 4. Top plate was fixed with the bottom plate by using four screws.

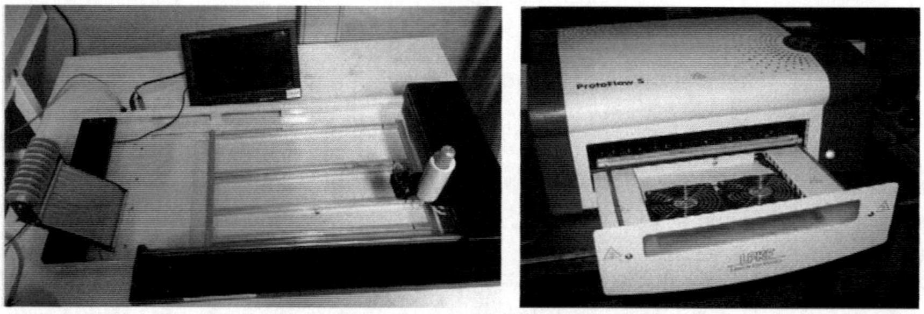

Figure 4. (a) Pick and Place Machine, (b) Reflow Oven.

4. Discussion

First step of mounting resonator pins are to put solder paste to the bottom plate. In PCB production a screen or stencil printer are used. In our case a pump was used to apply solder paste on pre marked positions. After that a manual pick and place machine was used and the pins were placed in the required position by using a alignment camera. Fig 5 shows the image of a groove resonator manufactured by using the pick and place machine. It has been observed that the reflow temperature can affect the pin position. If the reflow temperature is too high then the solder paste can start to flow and displace the pins. Different reflow profile has been checked and it has been observed that reflow profile mentioned in section 3 was most suitable for this device which gave better joint and better pin alignment. The reflow process is a combination of preheat, thermal soak, reflow, cooling stage. In our case the thickness of the bottom plate was 5mm, which was thicker than normal PCBs. However, it has been mentioned in [11] that the gap waveguide components can survive relatively large dimensional tolerance (compared to the conventional rectangular waveguide components) and ± 15μm of variation in pin position, pin height and pin size is not going to degrade the electrical performance of the gap waveguide components. This opens up the opportunity to use commercial pick and place assembly machines for fabricating gap waveguide based RF components.

Figure 5. Groove resonator manufactured by pick and place machine (a) without top plate (b) with top plate.

5. Conclusion

Gap waveguides is a new RF technology platform enabling RF system solutions to meet market demands. On the other hand conventional surface mount is ruling the electronic packaging world. In this paper we present a groove gap resonator, which is manufactured using a pick and place machine. This new technique is not only cost effective but time saving, versatile and easy to manufacture. A manual pick and place machine was used to assembly the resonator pins. After several trials it has been observed that this process is workable and the use of automatic pick and place machine can only improve the assembly quality.

Acknowledgements

This work was funded by Swedish Governmental Agency for Innovation Systems (VINNOVA) grant-2015-01387.

References

[1] Mounting of Surface Mount Components, National Semiconductor, August 2000.

[2] V. Jungnickel, K. Manolakis, W. Zirwas, B. Panzner, V. Braun, M. Lossow, *et al.*, "The role of small cells, coordinated multipoint, and massive MIMO in 5G," *Communications Magazine, IEEE,* vol. 52, pp. 44-51, 2014.

[3] J. Brady, N. Behdad, and A. M. Sayeed, "Beamspace MIMO for Millimeter-Wave Communications: System Architecture, Modeling, Analysis, and Measurements," *Antennas and Propagation, IEEE Transactions on,* vol. 61, pp. 3814-3827, 2013.

[4] F. Boccardi, R. W. Heath, A. Lozano, T. L. Marzetta, and P. Popovski, "Five disruptive technology directions for 5G," *Communications Magazine, IEEE,* vol. 52, pp. 74-80, 2014.

[5] D. Liu, B. Gaucher, U. Pfeiffer, and J.Grzyb, "Advanced Millimeter-Wave Technologies: Antennas, Packaging and Circuits,," *Chapter 6, John Wiley & Sons Ltd., ISBN: 9780470996171, 2009.*

[6] P. S. Kildal, "Three metamaterial-based gap waveguides between parallel metal plates for mm/submm waves," in *Antennas and Propagation, 2009. EuCAP 2009. 3rd European Conference on*, 2009, pp. 28-32.

[7] P. S. Kildal, E. Alfonso, A. Valero-Nogueira, and E. Rajo-Iglesias, "Local Metamaterial-Based Waveguides in Gaps Between Parallel Metal Plates," *Antennas and Wireless Propagation Letters, IEEE,* vol. 8, pp. 84-87, 2009.

[8] P. S. Kildal, A. U. Zaman, E. Rajo-Iglesias, E. Alfonso, and A. Valero-Nogueira, "Design and experimental verification of ridge gap waveguide in bed of nails for parallel-plate mode suppression," *Microwaves, Antennas & Propagation, IET,* vol. 5, pp. 262-270, 2011.

[9] Soldering Guidelines for Mounting Bottom-terminated Components, Application Note 62, Peregrine Semiconductor, 2016.

[10] S. Rahiminejad, A. U. Zaman, E. Pucci, H. Raza, V. Vassilev, S. Haasl, P. Lundgren,P.-S. Kildal, and P. Enoksson, *"Micromachined ridge gap waveguide and resonatorfor millimeter-wave applications,"* Sensors and Actuators A: Physical, vol. 186, pp. 264–269, 2012.

[11] A. U. Zaman and P.-S. Kildal, "GAP Waveguides," in *Handbook of Antenna Technologies*, Z. N. Chen, D. Liu, H. Nakano, X. Qing, and T. Zwick, Eds., ed Singapore: Springer Singapore, 2016, pp. 3273-3347.

IOP Conf. Series: Journal of Physics: Conf. Series **922** (2017) 012018 doi:10.1088/1742-6596/922/1/012018

Localized surface plasmon sensor using a hetero-core structured fiber optic with gold island films

Ai Hosoki*, Michiko Nisiyama, and Kazuhiro Watanabe

Faculty of Science and Engineering, Soka University, 1-236 Tangi, Hachioji, Tokyo 192-8577, Japan

*Corresponding author: ahosoki@soka.ac.jp

Abstract. This paper reports the effect of applying a gold islands film to the hetero-core structured optical fiber, which is fabricated by annealing thin Au films of 3 nm and 5 nm thicknesses and its sensing performance for the RI changes. We experimentally observe that novel LSPR spectra for 1.333 RIU appear in the visible-to-near-infrared region depending on the shape of Au islands. The absorbance and spectral sensitivities for a given refractive index region of the tested solvents were obtained to be 4.81 AU/RIU and 517 nm/RIU, respectively, in the range of 1.333–1.384 RIU in the case of 5-nm-thick Au film annealed at 900°C.

1. Introduction

Recently, localized surface plasmon resonance (LSPR) [1, 2] has been proposed as a way to minimize the size and cost of SPR sensors. A localized surface plasmon is generated by a light wave trapped within metallic nanoparticles (NPs) smaller than the wavelength of light. This effect leads to strong light absorption and scattering, in the appearance of surface plasmon absorption bands, for which the resonant frequency strongly depends upon the size and composition of the NPs. Therefore, a large number of optical fiber-based LSPR sensors for chemical sensing have been developed, including unclad [3] and tapered [4] optical fibers, which have advantages in terms of thin size, immunity to electromagnetic interference, and resistance to corrosion. The majority of these optical fiber-based LSPR sensors were fabricated by immobilizing Au NPs on the decladed fiber core surface by means of organic linkage [5]. However, these films are often non-uniform, with many agglomerates of deposited NPs for which it is difficult to control the density; this leads to poor stability and reproducibility of fiber surfaces and measurements.

Au island films are another possible way to produce plasmonic sensors with LSPR ability. As compared to Au NPs, Au island films are directly contact with the substrates and are easy to fabricate. Therefore, a number of studies have been reported to reveal the influence of such simple thermal treatments on the morphology of the Au layer evaporated on the glass substrates, as well as on its biosensing performance. Gupta et al. [6] revealed that for thinner Au films and higher annealing temperatures, the shape of the gold islands should be approximately spherical, with the shape becoming flatter at lower annealing temperatures. In order to construct LSPR sensors optical fiber-based LSPR sensors, Au island films have been applied to the unclad fiber [7] or the end of the fiber [8, 9]. Meriaudeau et al. [7] have firstly demonstrated the LSPR sensor effect in Au islands with an extra dielectric layer based on an unclad fiber. Beyond this, they proposed an LSPR sensor using Au island films deposited on the end of the fiber, and measured the absorption spectra for RI changes.

However, the ends of the fiber must be polished as its light intensity is affected by the end-face angle, resulting in a lower absorbance peak level for RI changes. Moreover, morphology and absorption spectra of the annealed Au layer coated on the unclad fiber surface has been rather limited, being mainly subject to sensor sensitivities for refractive index changes [10].

In this paper, we have reported the effect of applying a Au islands film to the cylindrical cladding surface, which is fabricated by annealing thin Au films of 3 nm and 5 nm thicknesses coated around the hetero-core structured optical fiber at temperatures of 500°C and 900°C, and have obtained the absorption spectra for the RI changes. Ever since the hetero-core technique [11] has been demonstrated by this paper's authors, a variety of hetero-core structured fiber optic SPR sensors for refractive index measurement and gas detection have been developed, owing to this technique's practical advantages in terms of simplicity of structure and its interesting sensing principle that requires no removal of cladding. Its structure consists of a short segment of single-mode optical fiber inserted between two multimode fibers. Because of the difference between the diameters of the fiber cores, the transmitted light is guided to the cladding region of the single-mode fiber, which works as a sensing interface. Most of the leaked light travels as an evanescent wave and can interact with metallic NPs on the cladding surface. The SEM images of Au island films on the fiber surface are obtained to estimate the size, shape, and density variations of the Au islands.

2. Sensor fabrication and operation principle

The fiber sensor used in this experiment is made with a hetero-core fiber structure, as shown in Fig. 1, which has previously been reported for use in SPR sensors [12, 13]. As mentioned above, this structure consists of an inserted fiber segment that works as a sensing region, the core diameter of which is smaller than that of the transmission fiber. The hetero-core sensor used here consists of an inserted single-mode (SI) fiber, with a 15 mm length and 3.1 μm core diameter, and spliced multimode (GI) fibers with core diameters of 50 μm at both ends. The outer diameters of both fibers are 125 μm, and the refractive index of the SI fiber used as a sensing surface is 1.4613. The SI sensor element is inserted by means of a thermal-fusion splicer (FSM-100M, Fujikura Ltd.). In this structure, due to the core diameter difference, most of the light propagated in the multimode fibers leaks into the cladding of the SI sensor element. An evanescent wave is generated at the surface of the cladding layer of the SI fiber when it is bounced off the boundary between the cladding region and the surroundings under a condition of total internal reflection. At the other end of the SI sensor element, some lights are re-coupled to the core of the multimode fiber. When metallic NPs are deposited onto the surface of the sensing region, they interact with the evanescent wave, resulting in a decrease of the reflection light corresponding to the RI of the surrounding medium.

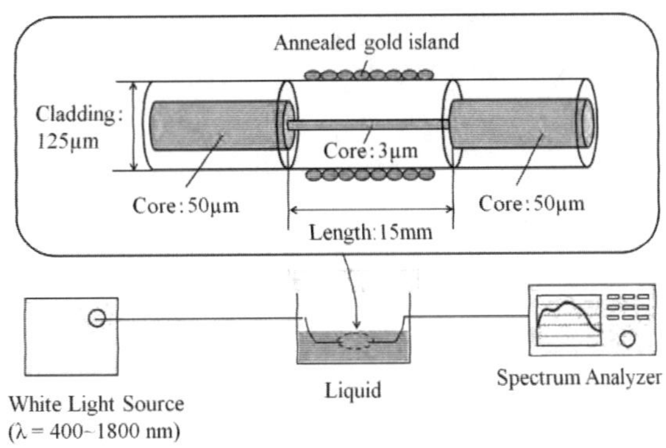

Figure 1. A hetero-core structured fiber optic LSPR sensor and experimental setup to analyze the response of gold island films fabricated on the hetero-core fiber to RI changes.

Thin Au films of thicknesses of 3 and 5 nm were cylindrically coated around the hetero-core region using an RF sputtering machine (CFS-4ES-231, Shibaura Mechatronics Corp.), in which a specially designed rotating mechanism was devised to achieve symmetrically uniform deposition on the cladding surface. The Au film's thickness is determined by the thickness of the deposited film on a glass substrate. After the film coating, the fibers were annealed at temperatures of 500°C and 900°C in an argon atmosphere for 1 h.

To characterize the propagated light spectrum of the LSPR sensors, a white light source, which emits continuous light waves in the range 400-1800 nm, and a spectrum analyzer (AQ6315A, Ando Electric Co., Ltd.) were used, as shown in Fig. 1. The hetero-core fiber structure was mounted on a stretching support, where the sensing portion was in a straight line. In the series of experiment, various glycerin solutions were prepared as test liquids with concentrations of 0%-40% by weight, corresponding to refractive indices ranging from 1.333 to 1.384. Spectrum measurement was carried out by immersing the entire sensing portion in the different RI solutions to be tested.

3. Results and Discussions

Figures. 2(a)–(d) show SEM images of 3- and 5-nm-thick Au films annealed at 500°C and 900°C. As shown in Figures. 2(a) and (b), 3-nm-thick Au annealed films form small particles with high island density and sizes (diameters) of 30 and 40 nm for 500°C and 900°C, respectively. In the case of annealed 5-nm-thick Au films, differently shaped islands are generated at the different annealing temperatures. In Figure. 2(c), the particles seem to be rather circular in shape with sizes of 60 to 80 nm for the case of 5-nm-thick Au films annealed at 500°C. On the other hand, Numerous irregular particles with non-circular shapes are seen on the entire surface of the 5-nm-thick Au film annealed at 900°C, as shown in Figure. 2(d). These results indicate that the sizes of the islands in the SEM images become larger than reported in the previous study by Gupta et al. [6], who found that circular particles ranged from 15 nm to 40 nm in diameter for Au-film thickness of 2 to 5 nm on a quartz substrate with an annealing time of 2 h. Therefore, it is estimated that the large size of the particles may stem from the annealing time effect.

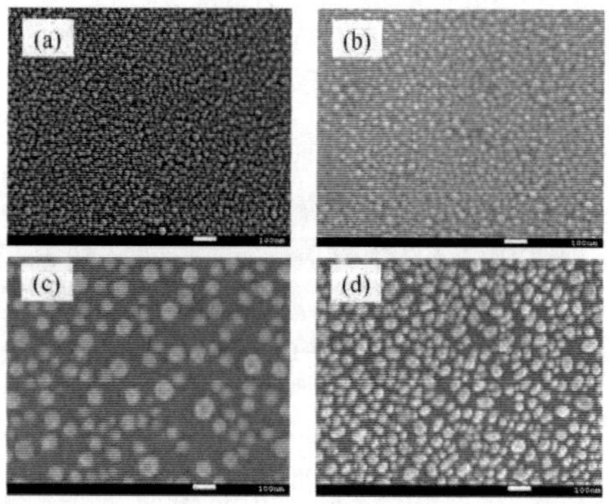

Figure 2. SEM images of the surface of the hetero-core fiber sensor based on gold-island films: (a) 3 nm-thick film annealed at 900 °C, (b) 3 nm-thick film annealed at 500 °C, (c) 5 nm-thick film annealed at 900 °C, and (d) 5 nm-thick film annealed at 500 °C..

The light intensity of each sensor in air is used as a reference signal, I_{ref}. The transmitted light intensity of each sensor through the liquid sample is I_t. The experimental absorbance ($-\log(I_t/I_{ref})$) is then obtained for various refractive index (1.333-1.384 RIU).

Figure. 3(a) and (b) shows the absorbance experimentally obtained for the 3-nm-thick Au film annealed at 900 °C, and 5-nm-thick Au film annealed at 900°C, as a function of wavelength for refractive indices of 1.333, 1.345, 1.358, 1.371, and 1.384, which correspond to glycerin concentrations of 0, 10.0, 20.3, 30.9, and 39.9 wt.%, respectively. It can be seen from Figure. 3(a) and

3(b) that as a result of the increasing refractive indices, the resonance wavelength is redshifted, and the light intensity decreases. The absorption spectra for all annealed sensors were similarly changed according to alternating glycerin concentrations. The absorbance at 627 nm and 720 nm was chosen to plot the LSPR response to the refractive index of glycerin solutions, as shown in Figure. 3(a) and (b).

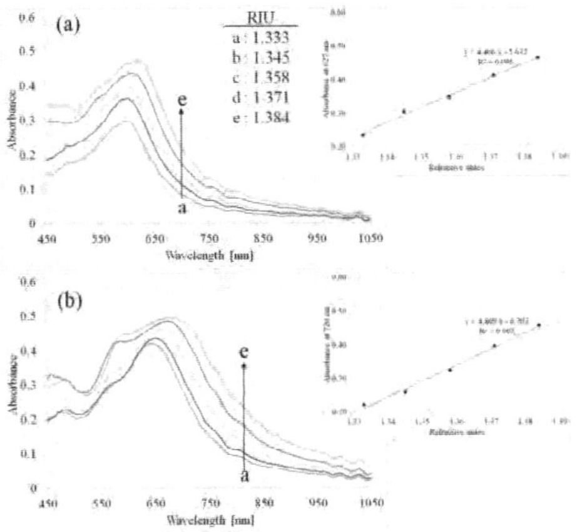

Figure 3. Experimental absorbance spectra for (a) 3 nm-thick Au films annealed at 900 °C and (b) 5 nm-thick gold films annealed at 900 °C for refractive index changes. The each inset shows the absorbance at 627 nm and 720 nm as a function of refractive index of glycerin solutions.

These wavelengths are selected in the plot because maximum sensitivities are found at those wavelengths. The sensitivity and linear correlation coefficient (R) are 4.41 absorbance unit (AU)/RIU (R =0.995) for annealed 5-nm-thick Au film and 4.81 AU/RIU (R = 0.993) for annealed 3-nm-thick Au film. These experimental results have proved that annealed sensors have high sensitivities compared to a conventional fiber-based LSPR sensor with Au-colloid film [14]. As can be obtained from the slopes in Figure. 4(a) and (b), the spectral sensitivities and limits of detection for the sensors with 3-nm-thick Au film annealed at 900 °C, and 5-nm-thick Au film annealed at 900 °C are found to be 385 nm/RIU and 1.3×10^{-3} RIU, and 517 nm/RIU and 9.7×10^{-4} RIU for the 1.333-1.384 RIU range, considering that the wavelength resolution of a spectral analyzer is 0.5 nm.

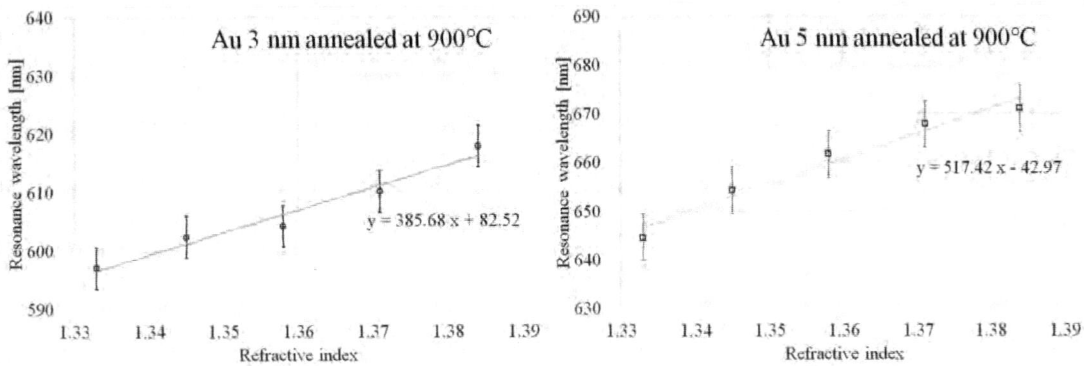

(a) (b)

Figure 4. Resonance shifts as functions of the refractive index for (a) 3 nm-thick gold film annealed at 900 °C and (b) 5 nm-thick gold film annealed at 900 °C

On the other hand, Figure. 5(a) shows that it is difficult to clearly read out the resonance peaks because of broad absorbance spectra in the cases of the 5-nm-thick Au sensor annealed at 500°C in the range of 1.333–1.384 RIU. However, refractive index measurements can be realized using a simpler scheme with a monochromatic light source and an optical power meter based on light-intensity-mode operation, because the obtained near-infrared spectrum region finds appreciable difference in their intensities as the given refractive index change. The tested sensor was found to have linear sensitivities of 2.93 AU/RIU for the 5-nm-thick Au sensor annealed at 500°C, as can be seen from Figure. 5(b).

The presence of an experimentally obtained LSPR absorbance peak is evidence that the evanescent field of the SI element interacts with the Au island film fabricated on the cylindrical cladding surface. In addition, the resonant wavelengths are redshifted and absorbance is linearly changed as the refractive index increases for all annealed Au film cases. The measured spectra show that it is possible to tune the absorption region in the case of an island film by controlling the shapes of the islands through Au-film thickness, annealing temperature, and annealing time during simple sensor fabrication.

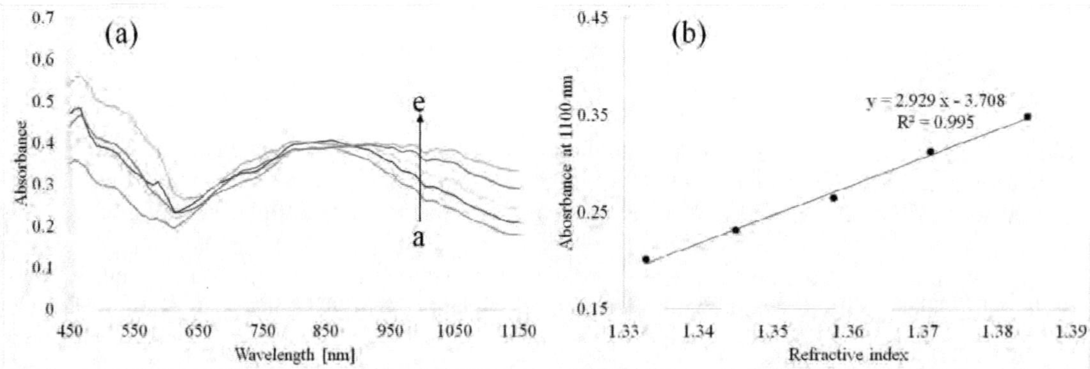

Figure 5. Experimentally obtained LSPR absorbance spectra for 5 nm-thick Au films annealed at (a) 700°C for RI changes (a: 1.333 - e: 1.384). Plot of absorbance at (b) 1100 nm for 5 nm-thick Au films annealed at 500°C as a function of refractive index of glycerin solutions.

4. Conclusion

In this paper, the effect of Au island films prepared by annealing thin Au films at high temperatures upon LSPR-based refractive index sensors has been experimentally demonstrated using a hetero-core structured optical fiber technique. The Au island films fabricated by annealing Au films with thickness of 3 and 5 nm at temperatures of 500°C and 900°C for 1 h were characterized. As a result, novel LSPR spectra for 1.333 RIU appeared in the visible-to-near-infrared region depending upon the shapes of the Au islands. The results indicated that for thinner Au films, the Au islands are hemispherical, whereas they are spheroidal for thicker films. In addition, lower annealing temperatures generate larger particles. The experimental results for refractive index measurements show that the resonant wavelengths were redshifted as the refractive index increased for all annealed Au film cases. Additionally, the 5-nm-thick Au film annealed at 900°C was found to be preferable from the viewpoint of sensitivity. The absorbance and spectral sensitivities for a given refractive index region of the tested solvents were obtained to be 4.81 AU/RIU and 517 nm/RIU , respectively, in the range of 1.333–1.384 RIU. This fiber-sensing structure will be attractive for practical sensor probes due to its sensitivity and simple sensor fabrication, and the controllability of its absorption region.

References

[1] K. M. Mayer and J. H. Hafner, "Localized Surface Plasmon Resonance Sensors," Chem. Rev. 111(6), 3828–3857 (2011).

[2] E. Petryayeva and U. J. Krull, "Localized surface plasmon resonance: Nanostructures, bioassays and biosensing– A review," Anal. Chim. Acta 706, 8–24 (2011).

[3] C. Ronot-Trioli, A. Trouillet, C. veillas, A. El-Shaikh, and H. Gagnaire, "Fiber optic chemical sensor based on surface plasmon monochromatic excitation," Anal. Chim. Acta 319, 121–127 (1996).

[4] C. L. Tien, H. W. Chen, W. F. Liu, S. S. Jyu, S. W. Lin, and Y. S. Lin, "Single-mode optical fibre surface plasma wave chemical sensor," Sens. Actuators B 41, 71–79 (1997).

[5] T. Okamoto, I. Yamaguchi, and T. Kobayashi, "Local plasmon sensor with gold colloid monolayers deposited upon glass substrates," Opt. Lett. **25**(6), 372–374 (2000).

[6] G. Gupta, D. Tanaka, Y. Ito, D. Shibata, M. Shimojo, K. Furuya, K. Mitsui, and K. Kajikawa, "Absorption spectroscopy of gold nanoisland films: optical and structural characterization," Nanotechnology 20(2), 025703_1–9 (2009).

[7] F. Meriaudeau, T. Downey, A. Wig, A. Passian, M. Buncick, and T. L. Ferrell, "Fiber optic sensor based on gold island plasmon resonance," Sens. Actuators B 54, 106–117 (1999).

[8] P. M. P. Gouvêa, H. Jang, I. C. S. Carvalho, M. Cremona, A. M. B. Braga, and M. Fokine, "Internal specular reflection from nanoparticle layers on the end face of optical fibers," J. Appl. Phys. 109, 103114 (2011).

[9] F. Meriaudeau, A. Wi, A. Passian, T. Downey, M. Buncick, and T. L. Ferrell, "Gold island fiber optic sensor for refractive index sensing," Sens. Actuators B 69, 51–57 (2000).

[10] I. Antohe, K. Schouteden, P. Goos, F. Delport, D. Spasic, and J. Lammertyn, "Thermal annealing of gold coated fiber optic surfaces for iproved plasmonic biosensing," Sens. Actuators B 229, 678-685 (2016).

[11] K. Watanabe, K. Tajima, and Y. Kubota, "Macrobending Characteristics of a Hetero-Core Splice Fiber Optic Sensor for Displacement and Liquid Detection," IEICE Trans. Electron. 83(3), 309–314 (2000).

[12] M. Iga, A. Seki, and K. Watanabe, "Hetero-core structured fiber optic surface plasmon resonance sensor with silver film," Sens. Actuators B 101, 368–372 (2004).

[13] A. Hosoki, M. Nishiyama, H. Igawa, A. Seki, Y. Choi, and K. Watanabe, "A surface plasmon resonance hydrogen sensor using Au / Ta_2O_5 / Pd multi-layers on hetero-core optical fiber structures," Sens. Actuators B 185, 53–58 (2013).

[14] S.F. Cheng and L.K. Chau, "Colloidal gold-modified optical fiber for chemical and biochemical sensing," Anal. Chem. **75**, 16-21 (2003).

Design of air blast pressure sensors based on miniature silicon membrane and piezoresistive gauges

J Riondet[1,2,3], A. Coustou[1,2], H. Aubert[1,2], P. Pons[1,2], M. Lavayssière[3], J. Luc[3], A. Lefrançois[3]

[1] CNRS-LAAS, 7 Avenue du colonel Roche, F-31400 Toulouse, France

[2] Université de Toulouse, LAAS-CNRS, F-31400 Toulouse, France

[3] CEA-DAM, F-46500 Gramat, France

Corresponding author : ppons@laas.fr

Abstract. Available commercial piezoelectric pressure sensors are not able to accurately reproduce the ultra-fast transient pressure occurring during an air blast experiment. In this communication a new pressure sensor prototype based on a miniature silicon membrane and piezoresistive gauges is reported for significantly improving the performances in terms of time response. Simulation results demonstrate the feasibility of a pressure transducer having a fundamental resonant frequency almost ten times greater than the commercial piezoelectric sensors one. The sensor uses a 5µm-thick SOI membrane and four P-type silicon gauges (doping level $\cong 10^{19}$ at/cm^3) in Wheatstone bridge configuration. To obtain a good trade-off between the fundamental mechanical resonant frequency and pressure sensitivity values, the typical dimension of the rectangular membrane is fixed to 30µm x 90µm with gauge dimension of 1µm x 5µm. The achieved simulated mechanical resonant frequency of these configuration is greater than 40MHz with a sensitivity of 0.04% per bar.

1. Introduction

The typical pressure over time during an explosion is shown in Figure 1 [1-2]. First of all, the pressure increases abruptly (with a rise time between 10 ns and 100 ns) from atmospheric pressure to reach the overpressure peak P_{max} (several tens of bar depending on the explosive load and the distance from the load). Then the pressure returns back to the atmospheric pressure during a positive phase in about 1 ms followed by a negative phase.

In order to validate the hydrocode, i.e. numerical simulations describing the shockwave discontinuity, an accurate measurement of the overpressure peak P_{max} is required [3], involving the use of pressure sensors presenting a short time response (<< 1µs) and then a high fundamental mechanical resonant frequency (>> MHz). Moreover, the high temperature environment during the explosion (> 1000 °C) makes the real-time dynamic pressure measurement of the blast very challenging [4-5].

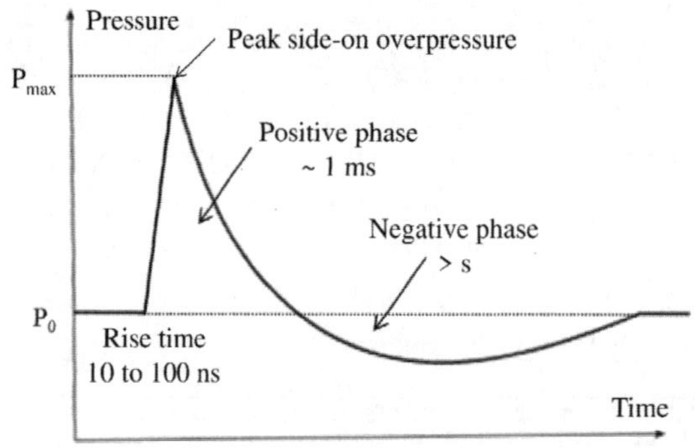

Figure 1: Typical time variation of the pressure during an air blast experiment

The commercial sensors used for the dynamic measurement of the pressure in harsh environment are usually piezoelectric pressure sensors or piezoresistive pressure sensors (Table 1). The fundamental resonant frequency of these sensors are generally lower than 1.5 MHz with maximum operating temperature lower than 65°C for sensors with the best dynamic response [6].

Table 1: Characteristics of available commercial sensors used for air blast experiments

Manufacturer	Reference	Technology	Sensing element	Pressure range (bar)	Temperature range (°C)	Fundamental resonant frequency (kHz)	Rise time (µs)
KISTLER [7]	211B	Piezoelectric	Quartz	3.5/ 35	-50 / 120	250 / 500	1 / 2
PCB [8]	113B	Piezoelectric	Quartz	3.5 / 70	-70 / 135	500 / 1500	< 1
PCB [9]	134A24	Piezoelectric	Tourmaline	70	0 / 50	1500	< 0.2
MULLER [10]	M60	Piezoelectric	PVDF	20	-20 / 65	2900	0.06
KULITE [11]	XTEH	Piezoresistive	Silicon gauge	3.5 / 70	-55 / 535	300 / 1000	
ENDVECO [12]	8540	Piezoresistive	Silicon gauge	3.5 / 35	-55 / 260	240 / 900	

Air blast experiments were performed at CEA-Gramat center using many commercial sensors mounted on pencil probes to measure the incident pressure, i.e., with sensor surface parallel to the direction of shock wave propagation [7-12] (Figure 2). A typical example of the response of such sensors is illustrated in Figure 3. It can be observed that the time response is too long to provide the accurate estimation of the overpressure peak P_{max}. The high cut-off frequency of piezoelectric sensors is only 20 % of the resonant frequency and the bandwidth is limited by the large dimensions of the sensing part (between 78 mm² and 450 mm²). Moreover typical piezoelectric sensors have a low cut-off frequency (> 0.5 Hz at -5 %) which is too high to follow accurately the overpressure decrease.

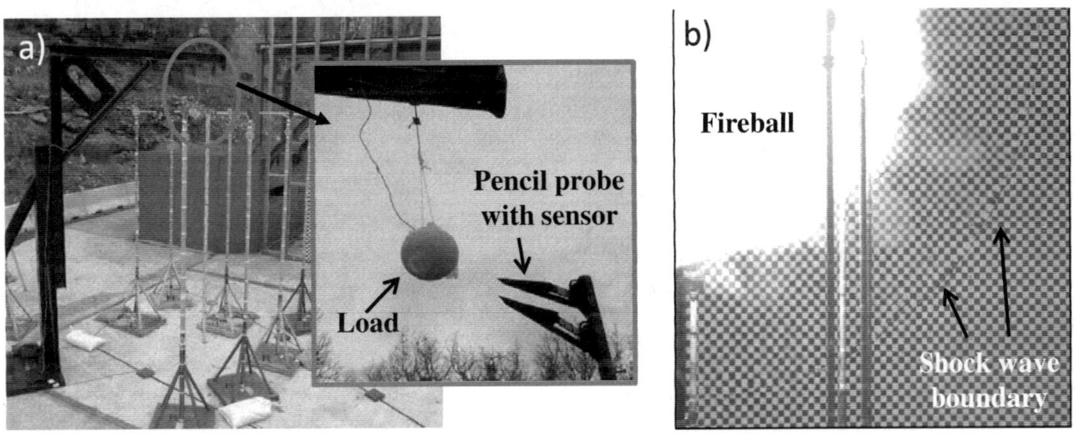

Figure 2: View of air blast experiment: (a) before the explosion and (b) 1.5 ms after the explosion

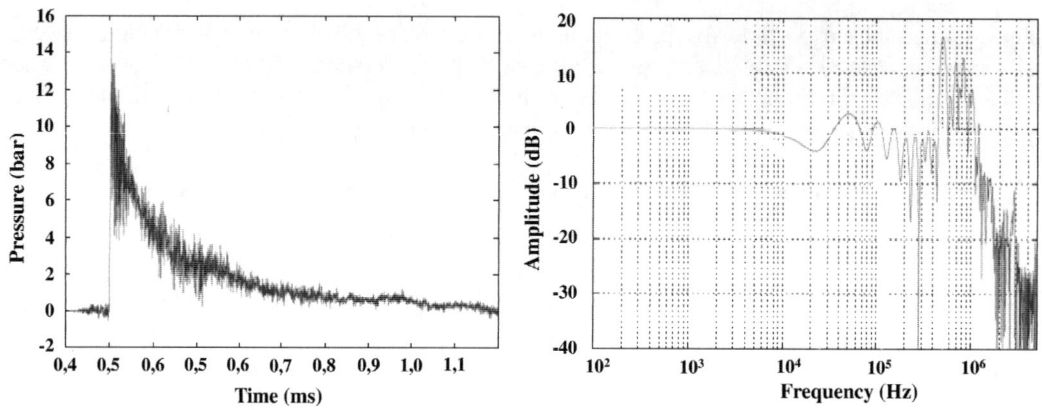

Figure 3. Typical measurement response of a pressure sensor during an air blast experiment in (a) time-domain and b) frequency-domain.

Fiber optic pressure sensors using Fabry–Perot cavities were studied in literature to increase the dynamic performances of air blast pressure sensors [13-16]. The principle is typically based on a thin dielectric membrane whose displacement modifies the Fabry-Perot cavity using an optical fiber. The best performances (4.1MHz of fundamental resonant frequency and a rise time of 120ns) were obtained in 2012 from a 3µm-thick silicon oxide membrane (125µm diameter) bonded one by one to the optical fiber by using a temperature reflow process [16]. Compared with the performances of available commercial sensors, the weak improvement on the dynamic response is obtained at the expense of the significant complexification of fabrication process.

The objective of this work is to achieve a device bandwidth at least 10 times greater than the bandwidth of the available commercial air blast pressure sensors. The targeted sensing element size is much smaller than $1mm^2$ in order to overcome the deterioration of the dynamic performances for incident pressure.
In order to fulfil the above technical requirements, we report here the design of a new piezoresistive pressure sensor based on a silicon membrane and silicon gauges. The piezoresitive detection is chosen here because it provides a better signal-to-noise ratio than their capacitive counterpart [17].

2. Piezoresistive sensor design

2.1. Sensor topology

The piezoresitive sensor combines (Figure 4) [18]:
- The N-type (100) mono-crystalline silicon membrane with edges parallel to <110> direction and fabricated from SOI (Silicon On Insulator) wafer;
- The P-type silicon gauge obtained from boron implantation. The doping level is of 10^{19} at/cm^3 to obtain a good trade-off between the sensitivity of the gauge to the temperature and to stresses. With length-over-width gauge ratio of 5 and for a gauge thickness of 300 nm, the gauge resistance is of few kΩ;
- The Wheatstone bridge. The gauges are located close to the edges of a square membrane or at the center of a rectangular membrane (see section 3).

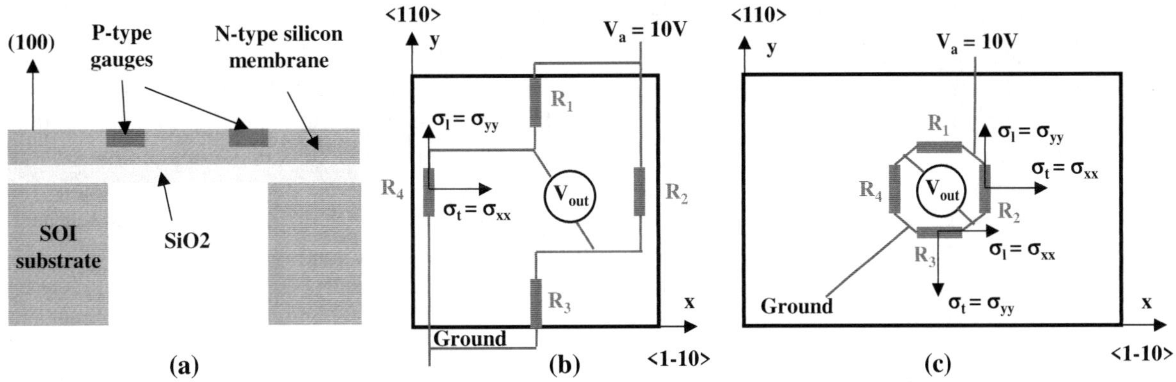

Figure 4: (a) Cross sectional view of the proposed piezoresistive sensor; (b) The Wheatstone bridge reported on a square membrane; (c) The Wheatstone bridge reported on a rectangular membrane

2.2. Mechanical resonant frequency of the designed piezoresistive sensor

The first technical requirement is the maximization of the fundamental mechanical resonant frequency of the silicon membrane. For rectangular membrane perfectly clamped at its edges, this frequency F_r is given by equation (1) [19] where E is the Young's modulus, v is the Poisson ratio, ρ is the membrane mass density, the dimensions e, a and b are respectively the thickness, width and length of the membrane and, α denotes a coefficient which depends on the aspect ratio b/a (see Table 2). It can be observed that the square membrane has a mechanical fundamental resonant frequency almost 60% higher than one achieved by rectangular membranes with large aspect ratio. Moreover rectangular membranes such that $b/a > 2$ have very close resonant frequencies. The aspect ratio is then fixed to 3 for stress uniformity considerations, as it will be shown in section 3.

The fundamental resonant frequency of square and rectangular ($b/a = 3$) membranes is displayed in Figure 5 as a function of the membrane width and for various membrane thicknesses. The minimum membrane width is here of 20 µm for technological reasons (aspect ratio of reactive ionic etching < 20). It can be observed that the design of a silicon membrane having a fundamental mechanical resonant frequency higher than several tens of MHz is possible.

$$F_r = \alpha * \frac{e}{a^2} * \sqrt{\frac{E}{\rho(1 - v^2)}} \qquad (1)$$

b/a	1	1.5	2	2.5	3	>>1
α	1.66	1.24	1.13	1.09	1.06	1.03

Table 2: Coefficient α as a function of the aspect ratio b/a

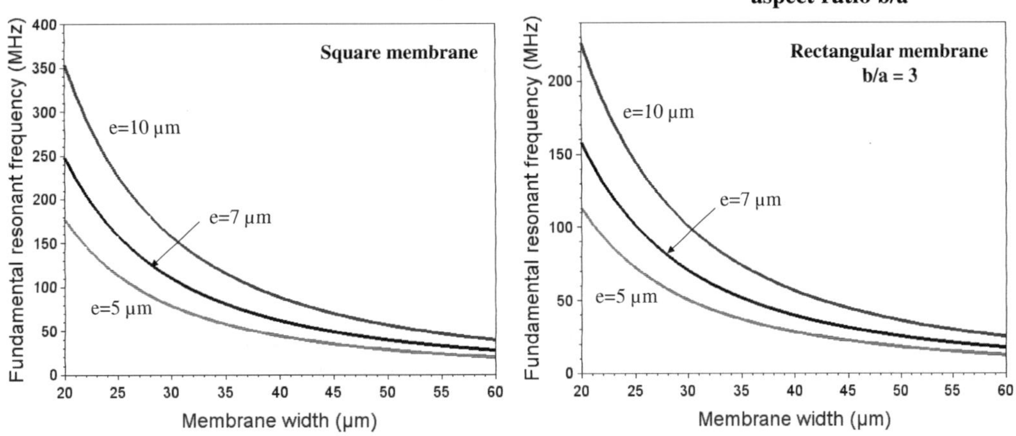

Figure 5: Fundamental mechanical resonant frequency of the membrane as a function of the membrane width for various membrane thicknesses

2.3. *Pressure sensitivity of the designed piezoresistive sensor*

When the four gauges of the Wheatstone bridge have the same initial resistance R_o and the same absolute sensitivity to pressure (that is, $\Delta R_1/R_o = \Delta R_3/R_o = -\Delta R_2/R_o = -\Delta R_4/R_o$), the voltage V_{out} at the bridge port is approximatively given by equation (2) [18] where V_a denotes the Wheastone bridge voltage alimentation, π_{44} is the main piezoresistive coefficient ($\pi_{44} \cong 10^{-9}$ Pa^{-1} for 10^{19} at/cm^3 P-type doping level) and σ_l (resp., σ_t) is the stress applied to the gauge parallel (resp., to the gauge perpendicular) to the current flowing through the gauge (see Figure 4). The stress at the membrane surface is derived from equation (3) [19], where P is the differential pressure between the two sides of the membrane and σ_n denotes a coefficient depending on the gauge position on the membrane and on the membrane geometry (Table 3). Assuming that the gauges are punctual, the sensor response can be predicted from equations (2)-(3) and from Table 3. Moreover, for the square membrane, we assume that the gauges are located at the centre of membrane edges and that the stress parallel to the membrane edge is negligible compared with other stresses. For rectangular membrane it is assumed that the gauges are located at the centre of the membrane and that the stress parallel to the membrane length (along the x-direction) is negligible. Figure 6 gives the voltage at the bridge port as a function of the membrane width for different membrane thicknesses and for the differential pressure P=1 bar. It can be observed that, even for the smallest membrane with a thickness of 5µm, the sensor sensitivity is higher than 0.03 % per bar. This is sufficiently high for the targeted application.

$$\frac{V_{Out}}{V_a} \cong \frac{\pi_{44}}{2}(\sigma_l - \sigma_t) \qquad (2)$$

$$\sigma = \sigma_n * \left(\frac{a}{e}\right)^2 * P \qquad (3)$$

Table 3: Coefficient σn as a function of the aspect ratio b/a and of the gauge location on the membrane surface

Aspect ratio b/a	1	1.4	1.8	≥ 2
Membrane edge (Stress perpendicular to the edge)	0.31	0.44	0.49	0.50
Membrane centre (Stress perpendicular to the membrane length)	0.14	0.21	0.24	0.25

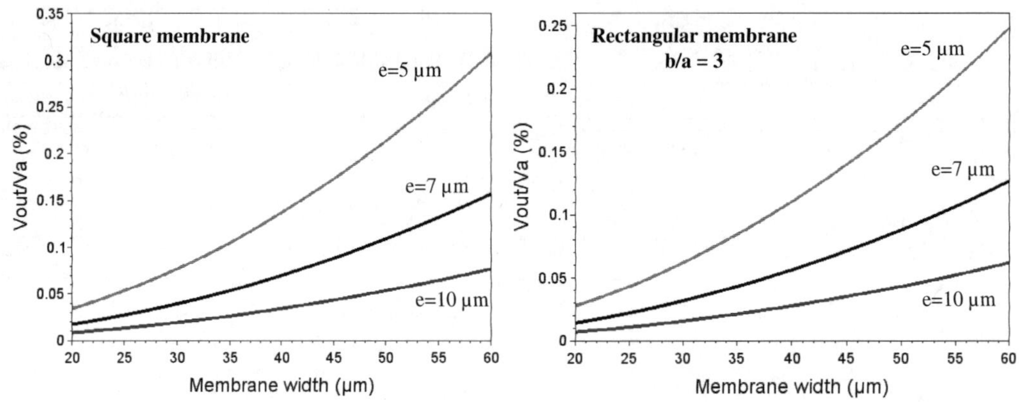

Figure 6: Normalized voltage at the bridge port versus the membrane width for a differential pressure of 1 bar. The voltage *Va* denotes the voltage that feeds the Wheastone bridge.

2.4. Measurement resolution

At high temperature (>1000°C), the electronic rms noise voltage V_{ng} at gauges ports in the gauges is dominated by the Johnson (or thermal) noise and is given by equation (4) where k_b (J/K) is the Boltzmann's constant, T (K) is the absolute temperature, R_0 (Ω) denotes the initial gauge resistance and Δf (Hz) is the bandwidth of the measurement.
In the Wheatstone bridge configuration shown in Figure 7, the unloaded output differential noise rms voltage V_{no} is given by equation (5). In practice, the bridge output port is connected to a load (R_L) through transmission line of characteristic impedance Z_C. The resulting loaded output differential noise rms voltage V_n is given in equation (6). As reported in Table 4, when $R_L = Z_C = 300\ \Omega$ the rms loaded differential noise voltage V_n is of 25 μV. For V_a =10 V and for a full scale response V_{out}/V_a = 1%, the measurement resolution is then lower than 0.03% of the full scale range.

$$V_{ng} = 2\sqrt{k_b T R_0 \Delta f} \qquad (4)$$

$$V_{no} = 8\sqrt{k_b T R_0 \Delta f} = 4\ V_{ng} \qquad (5)$$

$$V_n = 8\sqrt{k_b T R_0 \Delta f}\left[1 + \frac{1}{4}\sqrt{\frac{R_L}{R_0}}\right]\frac{R_L}{R_L + R_0} \qquad (6)$$

Table 4: Noise rms voltage

T	R_o	Δf	R_L=Z_C	V_{ng}	V_{no}	V_n
1000 °C	1000 Ω	35 MHz	300 Ω	50 μV	200 μV	25 μV

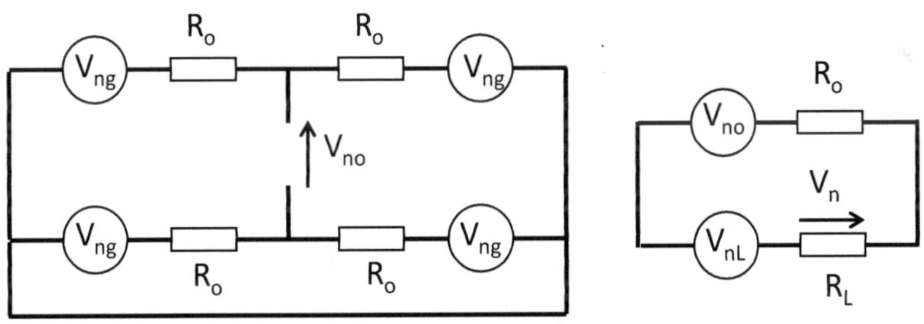

Figure 7: Equivalent noise circuit for the Wheatstone bridge for the unloaded (left) and loaded (right) configurations.

3. Piezoresistive sensor modelling

The sensors with perfectly clamped square and rectangular membranes were simulated using COMSOL software (Figure 8). The gauges were modelled by a 300 nm-thick P-type (10^{19} at/cm$^{3)}$ monocrystalline silicon and the interconnection by 5 μm-thick P-type (10^{20} at/cm^3) monocrystalline silicon. To obtain a good trade-off between the mechanical resonant frequency and pressure sensitivity values, the width of the square (respectively rectangular) membrane is fixed to 40 μm (respectively 30 μm). For the square membrane, the four gauges are not located exactly at the membrane edges in order to minimize the impact of eventual misalignment between gauges and membrane. For the rectangular membrane, the gauge location is constrained by the gauge length.

IOP Conf. Series: Journal of Physics: Conf. Series **922** (2017) 012019 doi:10.1088/1742-6596/922/1/012019

The stress distribution (σ_{xx} - σ_{yy}) is displayed in Figure 9 for square and rectangular membranes and for differential pressure of 1 bar. For square membrane, a fast variation of stress along the x and y directions occurs while for rectangular membrane, the stress distribution is quite uniform on the gauge surface, allowing more flexibility versus misalignment error between the gauges and membrane. Within these stress distributions, the sensor responses were calculated using Multiphysics COMSOL software. Table 5 summarized the sensors performances for the square and rectangular membranes. The resonant frequency may reach values between 30 MHz and 40 MHz, and the sensitivity is of 0.05% per bar.

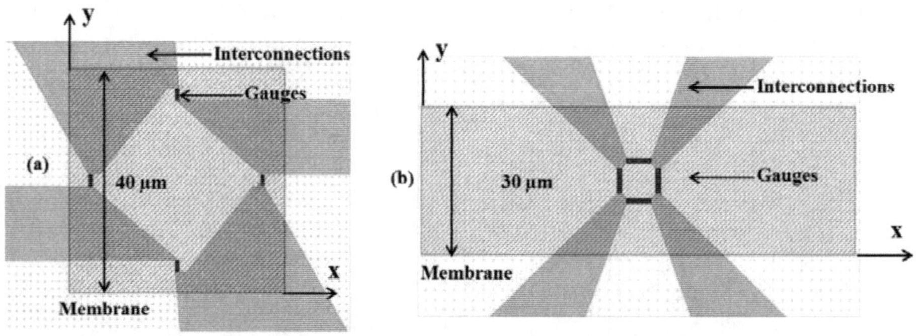

Figure 8: Sensor configurations simulated with COMSOL software with (a) a square membrane and (b) a rectangular membrane

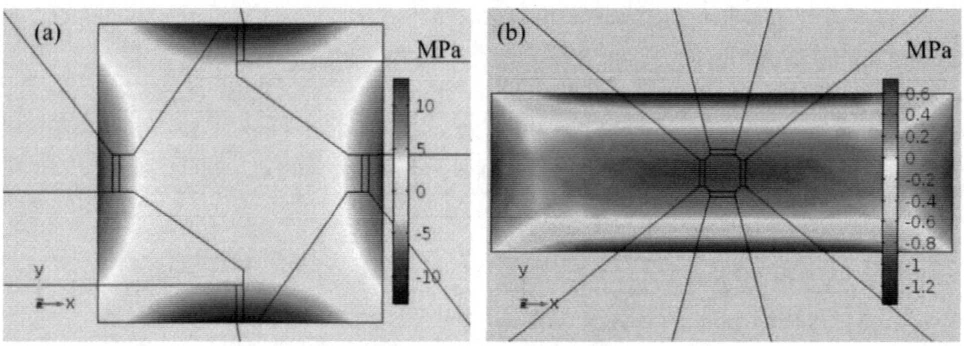

Figure 9: Stress (σ_{xx} -σ_{yy}) for a differential pressure of 1bar with (a) a square membrane and (b) a rectangular membrane

Table 5: Sensor dimensions and simulated performances

	Membrane thickness	Membrane Width Length	Gauge Width Length	Resonant frequency from Eq. (1)	Resonant frequency from COMSOL	Sensitivity from Eq. (2)	Sensitivity from COMSOL
Square membrane	5 µm	40µm 40µm	1µm 5µm	45 MHz	30 MHz	0.13% / bar	0.05% / bar
Rectangular membrane	5 µm	30µm 90µm	1µm 5µm	50 MHz	42 MHz	0.06% / bar	0.04% / bar

4. Conclusion

A miniaturized piezoresistive pressure sensor has been designed to overcome the limitations of available commercial piezoelectric sensors used for the monitoring of the dynamic pressure during air blast experiments. Modelling and simulation results show that it is possible to design sensors with a significant enhancement of the time response compared with the state-of-the-art.

Acknowledgements
This work was partially funded by Occitanie Region, France, through the COCNANO project.

References

[1] W. Fickett, C. William, Detonation: Theory and Experiment, Dover Publications, 2011.
[2] P.-L.Walter, Air-blast and the Science of Dynamic Pressure Measurements, Sound and Vibration, pp. 10-16, December 2004.
[3] P. L.Silver, Evaluation of Air Blast Measurement Techniques, 75th Shock and Vibration Proceedings, Virginia Beach, VA, October 17 - 22, 2004.
[4] [A. Lefrancois, G. Baudin, J.L. Cremoux, J. Massoni, R. Saurel, Blast Efficiency of Aluminized High Explosives, 17th International Symposium on Military Aspects of Blast and Shock, 2002
[5] L. Munier, E. Lapebie, C. Le Gallic, Secondary Combustion of High Explosive Detonation Products - Experimental Characterization of Thermal Effects, 20th International Symposium on Military Aspects of Blast and Shock, 2008
[6] F. Bauer, Ferroelectric Polymers for High Pressure and Shock Compression Sensors, Mat. Res. Soc. Symp. Proc., Materials Research Society, 698, 1-12, 2002
[7] https://www.kistler.com/by/en/search/?no_cache=1&tx_kesearch_pi1%5Bsword%5D=211B&tx_kesearch_pi1%5Bpage%5D=1&tx_kesearch_pi1%5BresetFilters%5D=0&tx_kesearch_pi1%5BsortByField%5D=&tx_kesearch_pi1%5BsortByDir%5D=
[8] http://www.pcb.com/Products.aspx?m=113B21 / 113B24 / 113B28
[9] https://www.pcb.com/products.aspx?m=134A24
[10] https://mueller-instruments.de/en/pressure-measurement/pressure-sensor-m/
[11] https://www.kulite.com/products/product-catalog/super-high-temperature-pressure-transducer-xteh-10l-190-xteh-10l-190m/
[12] https://buy.endevco.com/8540-pressure-5.html
[13] W. MacPherson et al., Blast-pressure measurement with a highbandwidth fibre optic pressure sensor, Meas. Sci. Technol. 11 (2000).
[14] S. Watson et al., Investigation of shock waves in explosive blasts using fibre optic pressure sensors, Meas. Sci. Technol. 17 (2006)
[15] W. Parkes et al., Design and fabrication of dielectric diaphragm pressure sensors for applications to shock wave measurement in air, J. Micromech. Microeng. 17 (2007)
[16] N. Wu et al., An ultra-fast fiber optic pressure sensor for blast event measurements, Meas. Sci. Technol. 23 (2012)
[17] E. Mile, Nanowire and carbon nanotube based NEMS resonator, PhD Thesis, June 2010
[18] M. Olszacki, Modeling and optimization of piezoresitive pressure sensors, PhD Thesis, July 2009
[19] R. J. Roark, W. C. Young, et R. G. Budynas, Roark's formulas for stress and strain, 7th ed. New York: McGraw-Hill, 2002

IOP Conf. Series: Journal of Physics: Conf. Series **922** (2017) 012020 doi:10.1088/1742-6596/922/1/012020

Development of a resonant asymmetric micro mirror using an electro active polymer actuation

F Casset[1, 2], P Poncet[1, 2], B Desloges[1, 2], B Neff[1, 2], F Domingues Dos Santos[3], JS Danel[1, 2], M Vimercati[4], L Zanotti[4] and S Fanget[1, 2]

[1]Université de Grenoble Alpes, F-38000 Grenoble, France
[2]CEA-LETI, MINATEC Campus, 38054 Grenoble, France
[3]ARKEMA-PIEZOTECH, F-69493 Pierre Benite, France
[4]STMicroelectronics, 20864 Agrate Brianza (MB), Italy

fabrice.casset@cea.fr

Abstract. In this paper we present the development of a resonant asymmetric micro mirror. In particular, we prove the possibility of using an Electro Active Polymer as actuation material. The working principle of the micro mirror and the design retained is first presented. Then the technological realization is summarized. We obtain micro mirror demonstrators and the first characterization results are given. These first promising results open the way of low cost and low temperature process actuators for Micro Electro Mechanical Systems.

1. Introduction

Micro Electro Mechanical System (MEMS) technologies lead to the development of miniaturized devices presenting high performances and low consumption [1]. In particular, micro mirrors are a key component for optical applications such as micro scanner, pico-projector or Light Detection And Ranging (LIDAR) systems. They consist in a movable electromechanical element, able to guide or deflect an optical beam or a laser. The actuation mechanism can be electrostatic [2], magnetic [3] or piezoelectric [4].

Among the various piezoelectric materials, Electro Active Polymers (EAPs), and in particular PVDF (Polyvinylidenefluoride) and derivatives fluorinated EAPs, are very interesting due to their ferroelectric properties, their low temperature processes and the possibility to be processed through standard printed electronic techniques [5]. They were already studied as actuator for example for haptic applications [6]. Our goal is to prove the possibility to use EAP as piezoelectric actuation material for a micro mirror.

In this paper, the work principle and the design of a resonant asymmetric micro mirror were presented. The main steps of the technological realization of demonstrator were also detailed. Finally, preliminary characterization results were given. The first promising measurement results proved the functioning of EAP actuators integrated in a silicon process flow. Further characterizations are in progress to evaluate the performances of this kind of innovative actuator for MOEMS devices.

2. Work principle and design

We aim to develop a resonant asymmetric micro mirror based on an Electro Active Polymer actuation.

As sketched in Figure 1, two piezoelectric actuator beams allow an out-of-plane displacement when an alternating actuation voltage is applied, thanks to the unimorph effect. It induces a rotation of the micro mirror due to the asymmetric position of the torsional springs which maintain the mirror.

The piezoelectric material is an Electro Active Pol-ymer (EAP) provided by Piezotech - Arkéma. PVDF (Polyinylidenefluoride) and derivatives fluorinated electroactive polymers are subject to numerous developments. Thanks to their ferroelectric semi-crystalline structures, they exhibit piezoelectric, pyroelectric or even relaxor ferroelectric properties that open new perspectives in sensors and actuators applications. Among them, P(VDF-TrFE) (Poly vinylidenfluoride-co-trifluoroethylene) copolymers are of particular interest since they crystallize directly from solution in a ferroelectric phase and can be processed through standard printed electronic techniques and using low temperature processes.

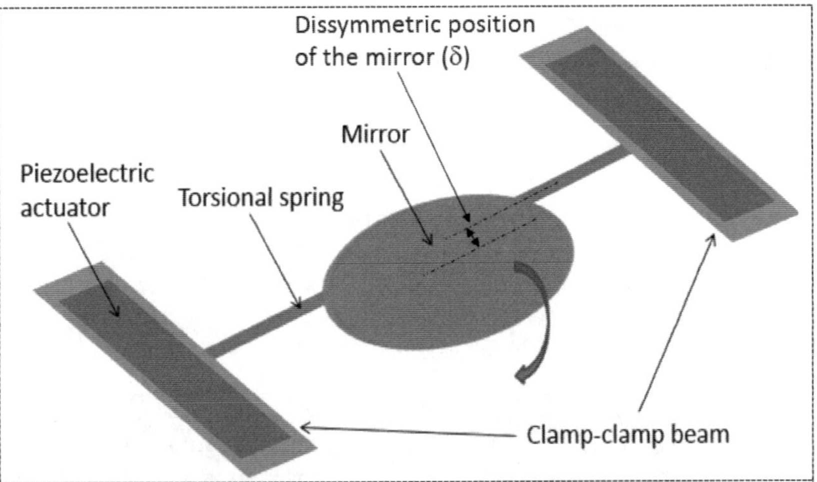

Figure 1. Resonant asymmetric micro mirror schematic view.

The design was performed using the Finite Element Method (FEM) approach and CoventorWare® software. Figure 2 gives a schematic view of the beam actuator. Based on previous work [7], we selected L_a=2/3 L_b. We chose to realize devices with L_b ranging from 3000 to 5000 µm which correspond to a theoretical resonant frequency ranging from 7.8 to 2.7 kHz. Then we adapt the actuator width (W_a) according to the beam width (W_b) to fulfill the EAP deposition requirements.

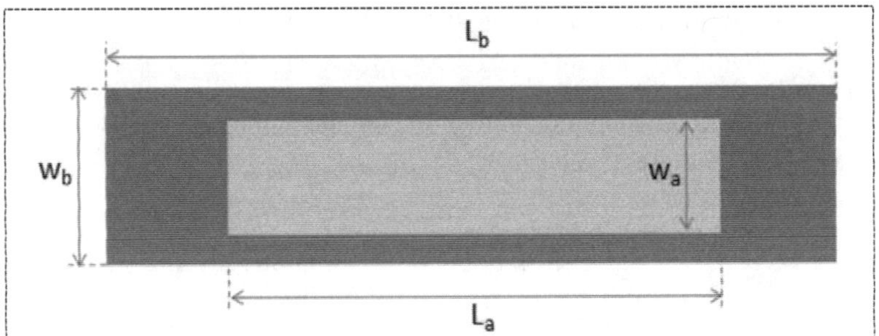

Figure 2. Beam actuator schematic view and main parameters.

Due to the asymmetric position of the mirror (δ), an angular rotation around torsional springs will result on the vertical displacement generated by the beam actuators. From [7], we select to study two different asymmetric positions of the mirror, δ = 50 or 100 µm.

The torsion beam design is one of the main challenges for efficient resonant asymmetric micro-mirror design. We know that the torsion mode efficiency is higher if the torsion beam width (w_t) is equal or smaller than the torsion beam thickness (h). Nevertheless, for technological reasons, we chose larger torsion beam widths, from 5 to 50 µm. The smaller w_t will allow to obtain the best rotation angle, but presents the highest technological risk.

The following table gives a synthesis of the main dimensions of our demonstrator.

Table 1. Resonant asymmetric micro mirror main dimensions.

Parameters	Dimensions (µm)
Mirror radius	250 - 1000
Beam length	3000 - 5000
Beam width	500
Torsion beam length	400 - 500
Torsion beam width	5 – 50
Thickness of the device	7.8

Using CoventorWare® and an arbitrary actuation voltage of 50 V, we obtain a mirror deflection of 48 µm (the lowest point at the end of the mirror) through harmonic simulations. It corresponds approximately to a rotation angle of about 5° using low actuation voltage. The simulated resonant frequency of the micro mirror with an actuation beam length of 3000 µm is 7.8 kHz, thus in good agreement with the theoretical resonant frequency of the single actuator beam.

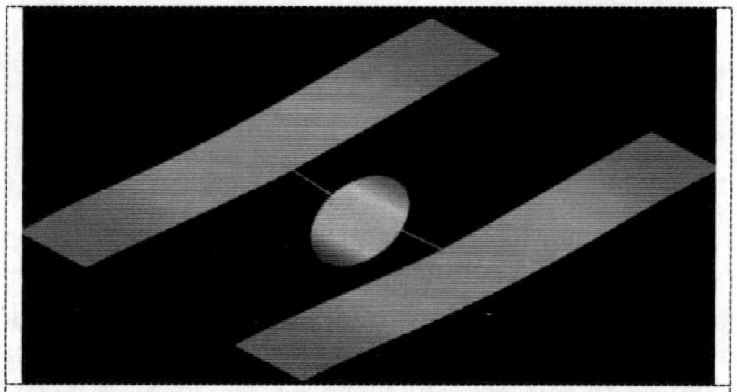

Figure 3. Beam Resonant asymmetric micro mirror deformation under 50V using FEM approach.

3. Technological realization

Figure 4 gives a schematic cross section of the device. Demonstrators were built using standard 8 inches silicon wafers and facilities.

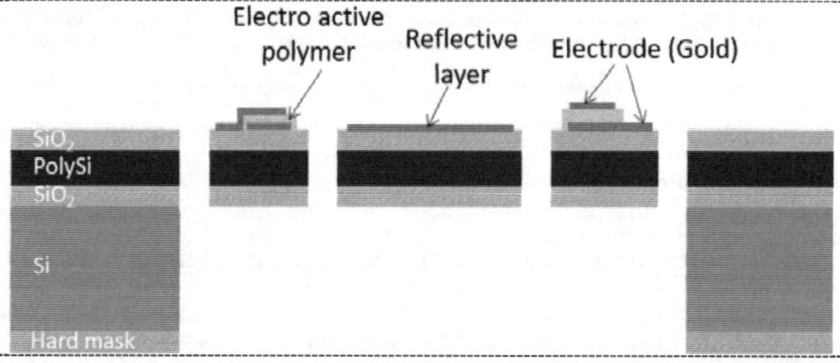

Figure 4. Schematic cross section of the resonant asymmetric micro mirror based on Electro Active Polymer actuation.

We first process the back side of the wafers to deposit and pattern the SiO_2 hard mask which will allow to release the micro mirror at the end of the process. Then SiO_2 and polysilicon layers were deposited in the front side of the wafer as structural layers and then etched to define the geometry of the MOEMS. A gold bottom electrode was deposited. This material is also used as reflective material, more suitable in IR spectral region, and it was deposited on the micro mirror part. The reflective properties of this gold material were evaluated using a multi wavelength ellipsometer. The measurement results indicate that the gold material used in our process presents a high reflectance value, in agreement with reference gold material (Figure 5).

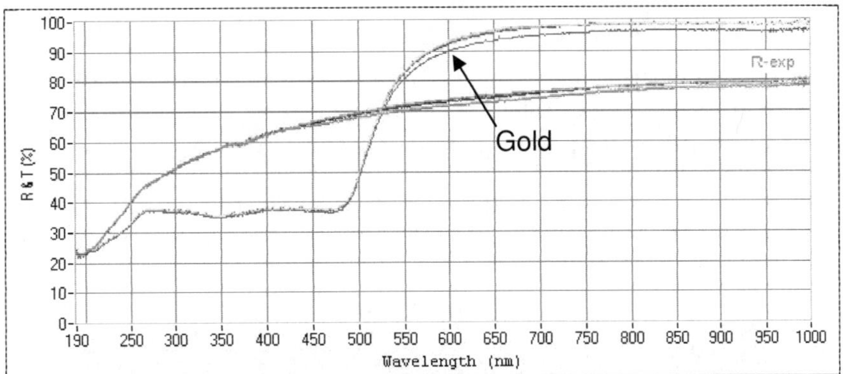

Figure 5. Reflective properties of the gold material above the micro mirror measurement using a multi wavelength ellipsometer, 9 points per wafer.

The EAP layer was deposited using screen printing technology and in particular an automatic Screen Printer EKRA X5 tool. The EAP material is a Piezotech - Arkéma FC Ink H 25 P(VDF-TrFE) copolymer solution in Cyclopentanone (viscosity 3 Pa.S). After each deposition step, the layer is annealed during 15 minutes at 135°C in an infrared oven. It permits to evaporate the solvent and crystallize the PVDF-TrFE film. We obtained an acceptable alignment of the EAP actuator on the silicon device (±50µm). Figure 6 gives a view of the silicon wafer with the EAP layer above the bottom electrode.

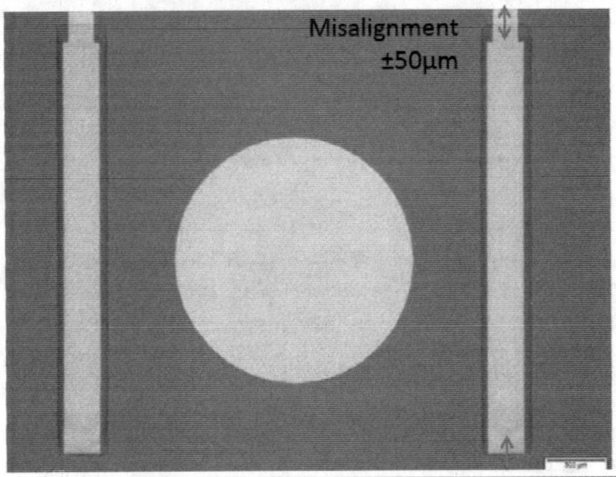

Figure 6. Photography of the resonant asymmetric micro mirror with the EAP layer deposited by screen printing above the bottom electrode. The EAP presents a misalignment of ±50 µm.

We measured the EAP actuator thickness and we observed a curved profile of the actuator. This profile is not problematic for the operation of our device but can be improved in the future, for example by adjusting the viscosity of the EAP solution (Figure 7).

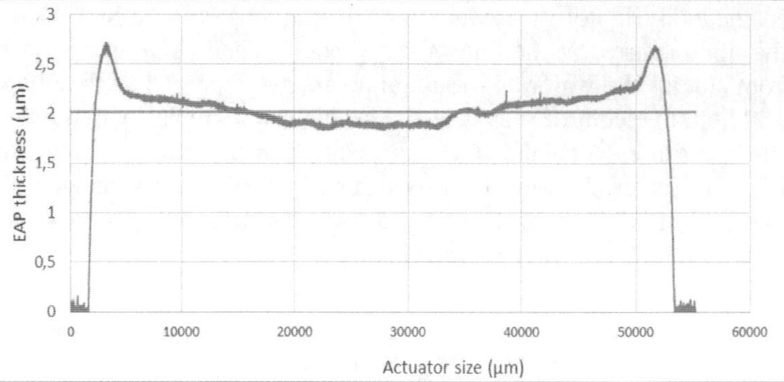

Figure 7. Photography Measurement using a mechanical profilometer of the profile of the EAP layer.

Top electrode and electrical lines in gold were then deposited on the top of the EAP layer, using a shadow mask. As shown in Figure 8, we obtain a quite good resolution and an acceptable alignment.

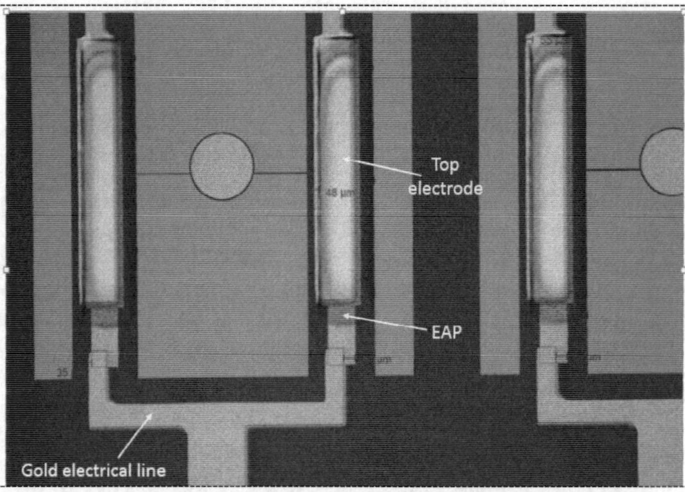

Figure 7. Photography of the resonant asymmetric micro mirror after gold deposition as top electrode, above the EAP, using a stencil.

Finally, the back side etching of the bulk silicon allows to release the MOEMS device can be seen in the Figure 9.

Figure 7. Photography of the resonant asymmetric micro mirror after back side etching of the bulk silicon. On the left the 200mm wafer and on the right a zoom on a single component.

4. Characterization

First, in order to activate the piezoelectric effect of the copolymer, actuators are poled with a square form signal (amplitude 160 V) at low frequency (1 Hz) during 120 s. This poling step induces the alignment of dipoles along the electrical field direction.

Then capacitance measurements were performed to check the EAP functioning, using an Agilent 4284A LRC meter. A capacitance value of 36 pF was measured on the actuators. It corresponds to a relative dielectric constant of 9, in good agreement with the expected value.

Electromechanical characterizations were also performed using a POLYTEC laser vibrometer MSA400. We used electrical probes to actuate our device as shown on Figure 10.

Figure 7. Photography of the MEMS micro mirror electromechanical characterization using a MSA400 laser vibrometer and electrical probes to actuate the device.

For this preliminary test, only 2 V_{RMS} were used to actuate the device. We focused on the first out-of-plane vibration mode of the beam actuator. A small dynamic deformation of the beam was measured (45 nm) using this low actuation voltage and it was unable to induce a notable vibration of the mirror. But a symmetric behaviour on the two actuators can be observed as expected (Figure 11). Higher actuation voltage should be used to actuate the mirror and evaluate the rotation angle which can be obtain using EAP actuators. Nevertheless, this first result proves the functioning of our innovative EAP actuator. The resonant frequency was measured at 15.97 kHz, thus higher than expected (7.8 kHz). It can be explained by residual stresses in the constitutive layers of the device.

Further measurements will be performed using higher actuation voltages to study our micro mirrors based on EAP actuators.

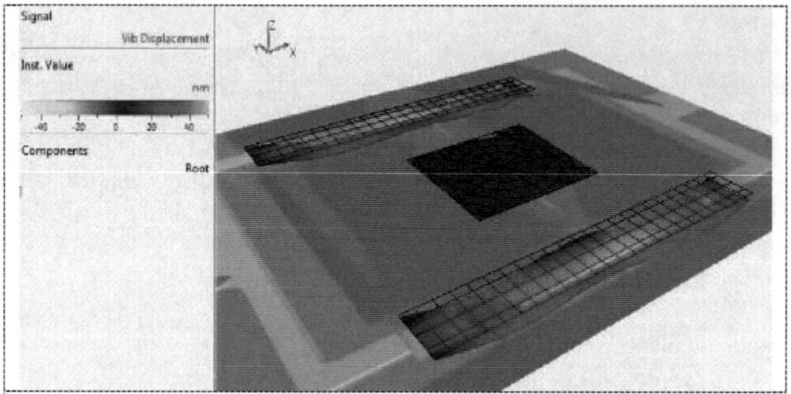

Figure 7. Laser vibrometer measurement on a micro mirror based on EAP actuators using an actuation voltage of 2VRMS.

5. Conclusion

In this paper the design of a resonant asymmetric micro mirror and the technological realization used to build demonstrator were presented. The electrical characterization of the actuators allows to extract a relative dielectric constant of 9, as expected. The preliminary electromechanical results prove the functioning of the actuators, which vibrate in their first out-of-plane resonant mode. Further characterization are in progress to obtain higher actuator displacement amplitudes, able to induce the rotation of the micro mirror.

Acknowledgments

The authors wish to acknowledge all the participants at the Lab4MEMSII project and the ENIAC funding, under grant agreement No 645145, to give us the opportunity to work on this innovative piezoelectric actuation material for silicon MOEMS.

References

[1] B. Vigna, "MEMS epiphany," presented at the IEEE 22nd Int. Conf. Microelectromech. Syst., Sor-rento, Italy, 2009.

[2] R. Farrugia, I. Grech, O. Casha, J. Micallef, E. Gatt, "Mechanical Optimisation of an Electrostatically actuated Scanning Mirror", WICT 2016, Workshop on ICT, June 2016.

[3] A.D. Yalcinkaya, H. Urey, D. Brown, T. Montague, R. Sprague, "Two-axis electromagnetic micro scanner for high resolution displays", IEEE/ASME Journal of Microelectromechanical Systems 15, 2006, pp.786–794.

[4] T. Bakke, P. Widdendorp, Z. Skokic, J. Gjessing, F. Tyhold, A. Vogl, "Applications of PZT-based micro mirrors", 5th International Workshop on Piezoelec-tric MEMS (PiezoMEMS), May 24-25, 2016.

[5] A. Plihon, V. Fischer, F. D. Dos Santos, and R. Gwoziecki, "Printed Actuators made with Electro-active Polymers on Flexible Substrates", 9th IEEE International Conference on Nano/Micro Engi-neered and Molecular Systems (NEMS), 2014, pp. 68–71.

[6] P. Poncet, F. Casset, A. Latour, F. Domingues Dos Santos, R. Gwoziecki, S. Fanget, " Development of haptic button based on electro active polymer ac-tuator", Proceedings of the 30th European Con-ference on sensors, actuators and microsystems (Eurosensors), Vol. 168, 2016, pp. 1500–1503.

[7] F. Filhol, "Micromiroirs piézoélectriques et micromi-roirs électrotatiques pour les grandes deviations de faisceaux lumineux", Thèse de l'université Joseph Fourier, soutenue le 19 novembre 2004.

IOP Conf. Series: Journal of Physics: Conf. Series **922** (2017) 012021 doi:10.1088/1742-6596/922/1/012021

Performance analysis of in-plane piezoelectric unimorph microactuators based on silicon and polymer substrates

Alex Díaz-Molina[1], **Jorge Hernando-García**[1] **and José Luis Sánchez-Rojas**[1]

[1]Microsystems, Actuators and Sensors Group, Universidad de Castilla-La Mancha, Ciudad Real, Spain.

alexdiazmolin@gmail.com

Abstract. This work describes the performance in terms of displacement of two different in-plane piezoelectrically-actuated microactuators: a gripper and an extensional actuator. The materials and the corresponding thicknesses were varied in search of the best configuration for maximum displacement. Three different combinations of materials were considered: AlN/silicon, AlN/polyimide and PVDF/polyimide. A static finite element analysis allowed the calculation of the in-plane displacement. Reduction in the displacement per volt was observed when the thickness of the structure was increased. Despite the fact that polymeric materials are significantly less stiff than silicon, having potential advantages, our simulations reveal that it does not translate into the corresponding improvement in displacement per volt.

1. Introduction

For many years, the creation of devices capable of operating in a reduced scale has been one of the most pursued challenges in the technological world. Advances in technology and fabrication methods has allowed producing devices in the micrometre scale known as MEMS (microelectromechanical systems). MEMS, based on mechanical parts and electronics components, combine low power consumption and high performance characteristics. Remarkable applications of MEMS has been achieved in different fields such as medicine, biology, robotics or fluidics [1-4].

Microactuators are an essential component of the former group. These devices have the potential of producing forces and displacement in their mobile parts, allowing them to interact with different objects. Several applications of microactuators required in-plane movement such as gripping, stroke amplifying or micropositioning [5-7].

In order to produce this movement, actuators can be driven by electrostatic, thermal, electromagnetic or piezoelectric [8-11] forces. We focus on piezoelectric materials, due to the possibility of implementing an all-electrical scheme, that allows for actuation and self-sensing. Common piezoelectric microactuators are silicon-supported, what results in high forces and low displacements. However nowadays there is also interest in the possibility of achieving very high displacements or low forces for applications such as soft materials assessment. And here is where other materials or thicknesses might play a new role [12-14].

The objective of this work is to study how the in-plane displacement of two types of actuators, an extensional actuator and a gripper, varies depending on the materials under use, as well as on the thickness of these materials. Here we compare three different combinations of materials: a rigid combination based on silicon as support and AlN as piezoelectric material; a soft combination based on PVDF and polyimide substrate; and a hybrid one combining AlN and polyimide. A Finite Element

IOP Conf. Series: Journal of Physics: Conf. Series **922** (2017) 012021 doi:10.1088/1742-6596/922/1/012021

model (FEM) allowed us to study the effect of the thickness of the different materials in the search for the maximum displacement.

2. Actuators Design

The case of study of this paper is based on two types of piezoelectric unimorph (also termed heterogeneous bimorph) actuators: symmetrical and asymmetrical. Figure 1 shows the cross section of the proposed structures. A symmetrical unimorph structure (depicted in figure 1(a)) consists of a substrate support with a piezoelectric layer between electrodes at each side of the substrate. The symmetrical design allows for in-plane and out-of-plane movements [15]. By using the electrodes configuration and the polarization direction shown in figure 1(a), in-plane movement can be achieved.

Asymmetrical unimorph structure (figure 1(b)) consists of a piezoelectric layer between electrodes joined to a substrate. Typically this configuration produces bending moment in the structure and, consequently, out-of-plane displacement. In-plane displacement can be implemented with this scheme by the proper design of the electrode layout [17].

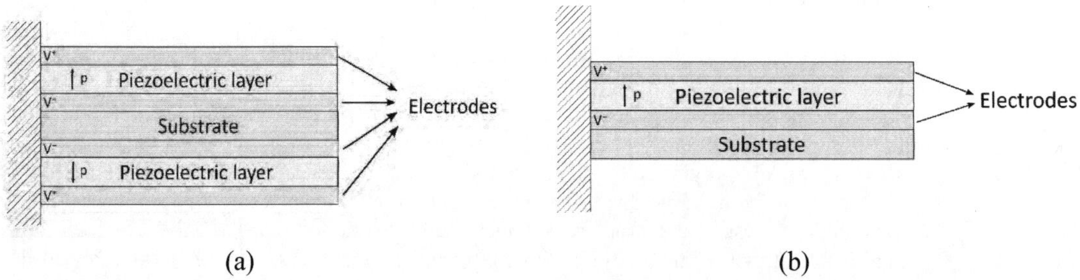

Figure 1. Schematic representation of unimorph structure: (a) symmetrical ; (b) asymmetrical

Two types of devices were considered: an extensional actuator with a symmetrical unimorph configuration, and a gripper with an asymmetrical unimorph structure. In order to maximize the displacement of the actuators, topology optimization was applied [16,17]. The optimization approach allows for the simultaneous design of both the shape of the structure and the electrode profile. The materials used for the optimization were silicon for the substrate and AlN for the piezoelectric layer with thicknesses of 100 μm and 1 μm, respectively. Figure 2 shows the shape and the electrode profile obtained after the optimization process for a 1000x1000 μm^2 extensional actuator and 5000x5000 μm^2 gripper. The different colours in the electrode profile represents different polarities in the voltage applied.

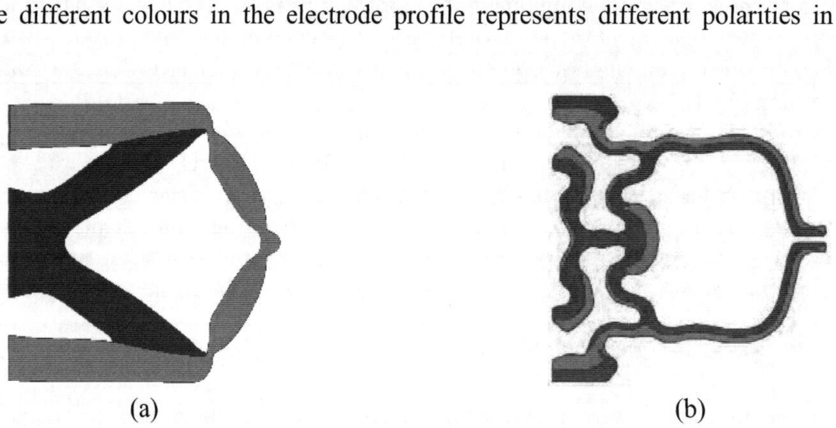

Figure 2. Schematic representation of the structure and electrode profile resulting from the optimization process: each colour means opposite voltage polarity (a) extensional actuator; (b) gripper

In the case of the extensional actuator, in-plane displacement can be obtained by covering the structure with just one electrode. Nevertheless, the optimization provides an electrode profile that maximizes the displacement. In the gripper model, and additional restriction was included in the optimization problem with the purpose of avoiding out-of-plane displacement in the tip. It is worth pointing out that these two device designs are the result of the optimization algorithm for the particular materials and dimensions mentioned above. Other materials or dimensions would require the corresponding optimization task and the resulting structure and electrode profiles would not be necessarily the same. Nevertheless, we use these particular designs as a reference to compare the effect of changing materials and dimensions on the microactuator performance.

3. FEM model description

A Finite Element Method (FEM) analysis has been conducted in order to study the tip displacement of the extensional model and the jaw behaviour in the gripper. Although the materials and the corresponding thicknesses were fixed for the optimization study, these parameters were modified during the FEM analysis. Three different sets of materials have been considered for the substrate and the piezoelectric layer of each model: AlN/silicon, AlN/polyimide and PVDF/polyimide. Table 1 shows the physical properties of the materials used in the simulations.

During the simulations, the structure was clamped in one of its sides (shown in figure 3). A Direct Current (DC) analysis was performed with 100V applied. As previously mentioned, each colour represents different polarities, being the blue areas 100V and the red ones -100V. Furthermore, the thickness of the electrodes was neglected in the FEM calculations. The displacement produced by the piezoelectric effect was measured along the X axis for the extensional actuator (figure 3(a)). In the case of the gripper, due to the symmetry in the displacement, only the motion of one of the arms was measured along the Y axis.

Table 1. Electromechanical characteristics of the materials

Material	Young Modulus (GPa)	Poisson ratio	Density (g/cm^3)	$d_{31}(pC/N)$
AlN	344	0.24	3.2	2
PVDF	2	0.34	1.76	19
Silicon	130	0.28	2.33	-
Polyimide	2.3	0.34	1.42	-

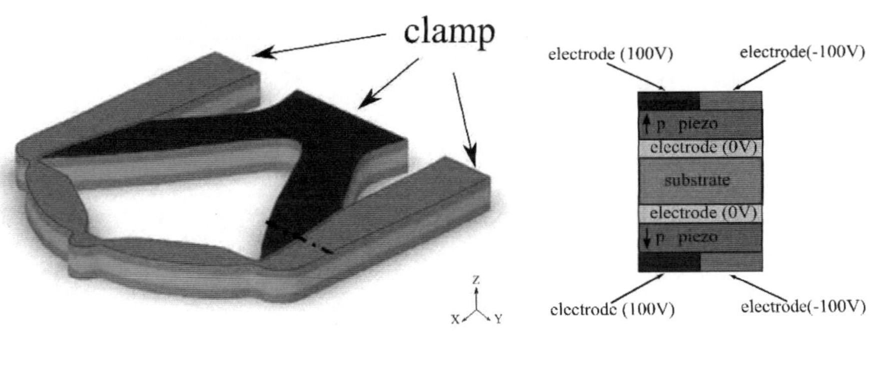

(a)

IOP Conf. Series: Journal of Physics: Conf. Series **922** (2017) 012021 doi:10.1088/1742-6596/922/1/012021

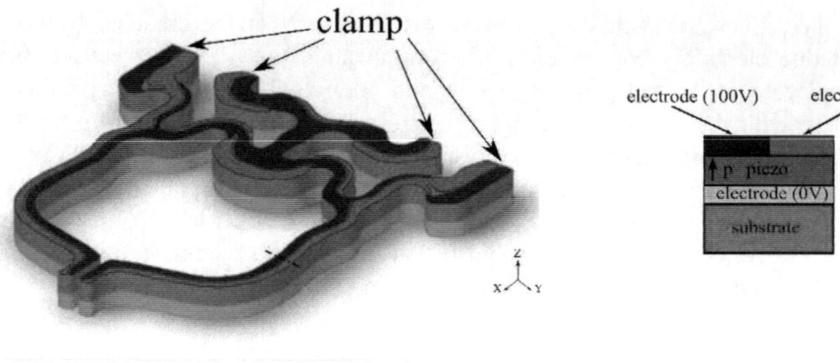

(b)

Figure 3. 3D model and inset of the simulated designs: (a) extensional actuator; (b) gripper

With the purpose of meshing the model, 3D element type has been employed. Hexahedral elements with squared faces and second order of interpolation were selected. A total number of 200,000 volume nodes were generated.

4. Simulation results

In the first analysis preformed, the thickness of the piezoelectric layer was fixed to 5 μm (in the case of the symmetrical model divided into two layers of 2.5 μm) and the thickness of the substrate was varied in the range from 0.04 μm to 800 μm. The displacements per volt obtained in absolute value for each model are presented in figures 4(a) and 4(b). By observing both graphics, it can be noticed that the greatest value of displacement per volt is obtained when the thickness of the substrate is minimum in all the cases proposed. If we pay attention to the combinations where AlN is present, the displacement tends towards the same value as the substrate thickness decreases, as expected due to the negligible role of the substrate; while for thicker substrates, the AlN/silicon combination results in the lower displacement due to the higher stiffness of silicon in comparison with polyimide. The soft model, formed by polyimide and PVDF, offers the largest displacement when the substrate is thin, due to the lower stiffness of PVDF; and its behaviour is similar to the rigid AlN/silicon combination when the thickness increases. The differences between both cases (rigid and soft) are greater in the gripper design than in the extensional model. The hybrid design, formed by AlN and polyimide, exhibits low variability in its motion in a wide range of thicknesses of the substrate. As it can be observed, the

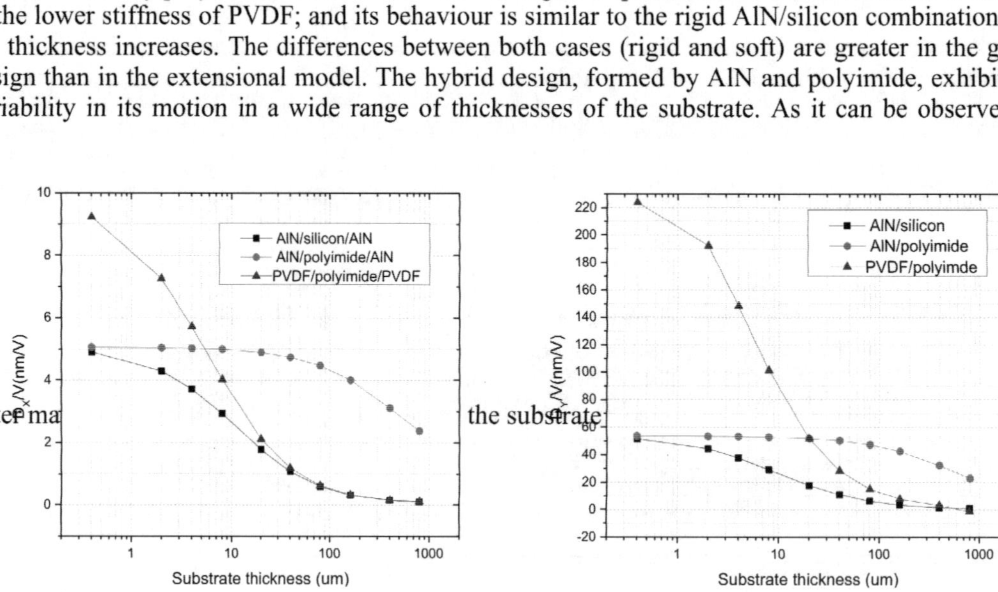

(a) (b)

Figure 4. Displacement per volt in the tip of the actuator as a function of the substrate thickness in a 5 μm piezoelectric layer: (a) extensional actuator; (b) gripper

latter ma ... the substrate

Variations on the thickness of the piezoelectric layer were also studied while the substrate was fixed to 8 μm. Figure 5 shows the behaviour of the displacement per volt in absolute value with varying thickness of the piezoelectric layer. As it can be observed, by increasing the thickness of the piezoelectric layer, the displacement of the tip decreases. In the case of the extensional design, the hybrid model exhibits the highest sensitivity to this parameter. For the gripper design it can be observed a similar behaviour.

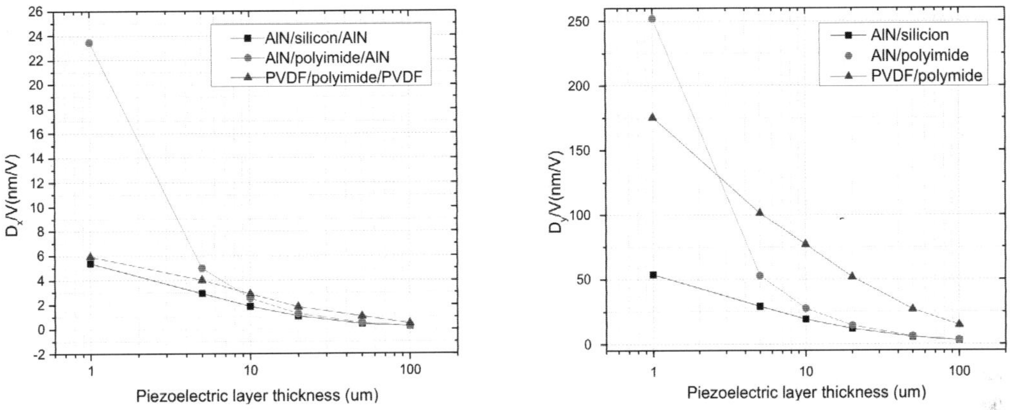

(a) (b)

Figure 5. Displacement per volt in the tip of the actuator as a function of the piezoelectric layer thickness in a 8 μm substrate layer: (a) extensional actuator; (b) gripper

5. Conclusion

A comparative study was carried out with different type of structures (symmetrical and asymmetrical), designs (extensional and gripper) and materials (AlN, PVDF, silicon and polyimide). The design of each model was obtained in a topological optimization process, with the purpose of maximizing the in-plane displacement. A static analysis was performed applying DC voltage in the electrodes and the resulting displacement produced in the tip was calculated for each model. Taking a topologically optimized layout as a reference, the thicknesses of the different layers and the materials were modified and its behaviour studied. A reduction in the displacement per volt was observed when the thickness of the structures was increased, independently of the layer varied. Even though polymeric materials exhibit significantly lower stiffness than silicon, an outstanding performance in terms of displacement per volt was not observed when they were used. Nevertheless, some materials combination suggest promising applications such as AlN and polyimide. This combination offers, for a wide range of substrate thickness, values of displacement per volt closer to the case where the thickness is negligible.

Acknowledgements
This work was supported by Spanish Ministerio de Economía y Competitividad project TEC2015-67470-P.

References

[1] Grant, S. A., Bettencourt, K., Krulevitch, P., Hamilton, J., & Glass, R. (2001). In vitro and in vivo measurements of fiber optic and electrochemical sensors to monitor brain tissue pH. Sensors and Actuators B: Chemical, 72(2), 174-179.
[2] Serrell, D. B., Law, J., Slifka, A. J., Mahajan, R. L., & Finch, D. S. (2008). A uniaxial bioMEMS device for imaging single cell response during quantitative force-displacement

measurements. Biomedical microdevices, 10(6), 883.

[3] Murthy, R., & Popa, D. O. (2009, January). Articulated MEMS Robot for Microfactory Applications. In ASME 2009 International Design Engineering Technical Conferences and Computers and Information in Engineering Conference (pp. 777-783). American Society of Mechanical Engineers.

[4] Kohl, M., Skrobanek, K. D., & Miyazaki, S. (1999). Development of stress-optimised shape memory microvalves. Sensors and Actuators A: Physical, 72(3), 243-250.

[5] Sheikh, S. A., & Shanmuganantham, T. (2014, January). A novel design microgripper based on electrothermal expansion principle. In Computer Communication and Informatics (ICCCI), 2014 International Conference on (pp. 1-5). IEEE.

[6] Conway, N. J., & Kim, S. G. (2004). Large-strain, piezoelectric, in-plane micro-actuator. In Micro Electro Mechanical Systems, 2004. 17th IEEE International Conference on.(MEMS) (pp. 454-457). IEEE.

[7] Lau, G. K., Yang, J., Tan, C. P., & Chong, N. B. (2016). An electro-thermally activated rotary micro-positioner for slider-level dual-stage positioning in hard disk drives. Journal of Micromechanics and Microengineering, 26(3), 035016.

[8] Eltagoury, Y. M., Soliman, M., Sabry, Y. M., Alotaibi, M. J., & Khalil, D. (2016). Electrostatic Comb-Drive Actuator with High In-Plane Translational Velocity. Micromachines, 7(10), 188.

[9] Meszmer, P., Hiller, K., Hartmann, S., Shaporin, A., May, D., Rodriguez, R. D., ... & Wunderle, B. (2014). Numerical characterization and experimental verification of an in-plane MEMS-actuator with thin-film aluminum heater. Microsystem Technologies, 20(6), 1041-1050.

[10] Ahmed, M. G., Dennis, J. O., Khir, M. H., Rabih, A., & Mian, M. U. (2016, November). Characterization of PolyMUMPs-based in-plane electromagnetic actuator. In A. H. Bhat, N. B. Yahya, A. B. Ramli, H. Soleimani, S. B. Zakariah@ Zakaria, & I. Faye (Eds.), AIP Conference Proceedings (Vol. 1787, No. 1, p. 050014). AIP Publishing.

[11] Nadig, S., Ardanuc, S., & Lal, A. (2013, July). Planar laser-micro machined bulk PZT bimorph For in-plane actuation. In Applications of Ferroelectric and Workshop on the Piezoresponse Force Microscopy (ISAF/PFM), 2013 IEEE International Symposium on the (pp. 152-155). IEEE.

[12] Bienaimé, A., Chalvet, V., Clévy, C., Gauthier-Manuel, L., Baron, T., & Rakotondrabe, M. (2015). Static/dynamic trade-off performance of PZT thick film micro-actuators. Journal of Micromechanics and Microengineering, 25(7), 075017.

[13] Fu, Y., Harvey, E. C., Ghantasala, M. K., & Spinks, G. M. (2005). Design, fabrication and testing of piezoelectric polymer PVDF microactuators. Smart materials and structures, 15(1), S141.

[14] Mastronardi, V. M., Guido, F., Amato, M., De Vittorio, M., & Petroni, S. (2014). Piezoelectric ultrasonic transducer based on flexible AlN. Microelectronic Engineering, 121, 59-63.

[15] Rogacheva, N. N., Chou, C. C., & Chang, S. H. (1998). Electromechanical analysis of a symmetric piezoelectric/elastic laminate structure: theory and experiment. IEEE transactions on ultrasonics, ferroelectrics, and frequency control, 45(2), 285-294.

[16] Ruiz, D., Donoso, A., Bellido, J. C., Kucera, M., Schmid, U., & Sánchez-Rojas, J. L. (2016). Design of piezoelectric microtransducers based on the topology optimization method. Microsystem Technologies, 22(7), 1733-1740.

[17] Ruiz, D., Díaz-Molina, A., Sigmund, O., Donoso, A., Bellido, J. C., & Sánchez-Rojas, J. L. (2017) Optimal design of a microgripper-type actuator based on AlN/Si heterogeneous bimorph. SPIE Microtechnologies 2017 conference. Paper 10246-4

Wafer-scale growth of highly textured piezoelectric thin films by pulsed laser deposition for micro-scale sensors and actuators

M. D. Nguyen[1,2,*], R. Tiggelaar[1], T. Aukes[2], G. Rijnders[1] and G. Roelof[1]

[1] MESA+ Institute for Nanotechnology, University of Twente, P.O. Box 217, 7500AE Enschede, The Netherlands

[2] Solmates B.V., Drienerlolaan 5, 7522NB, Enschede, The Netherlands

* E-mail: d.m.nguyen@utwente.nl

Abstract. Piezoelectric lead-zirconate-titanate (PZT) thin films were deposited on 4-inch (111)Pt/Ti/SiO_2/Si(001) wafers using large-area pulsed laser deposition (PLD). This study was focused on the homogeneity in film thickness, microstructure, ferroelectric and piezoelectric properties of PZT thin films. The results indicated that the highly textured (001)-oriented PZT thin films with wafer-scale thickness homogeneity (990 nm ± 0.8%) were obtained. The films were fabricated into piezoelectric cantilevers through a MEMS microfabrication process. The measured longitudinal piezoelectric coefficient (d_{33f} = 210 pm/V ± 1.6%) and piezoelectric transverse coefficient (e_{31f} = -18.8 C/m^2 ± 2.8%) were high and homogeneity across wafers. The high piezoelectric properties on Si wafers will extend industrial application of PZT thin films and further development of piezoMEMS.

1. Introduction

Pulsed laser deposition (PLD) differs from other physical vapour deposition techniques because of the very high kinetic energy of the ablated particles arriving at the substrate and the high deposition rate during each laser pulse. Already since 1965, Smith and Turner utilized a pulsed ruby laser to deposit the first vacuum deposited thin films [1], the PLD technique is only widely known when Dijkkamp *et al.* [2] were able to laser-deposit a thin film of $YBa_2Cu_3O_{7-\delta}$, a high-temperature superconductor material, which was of superior quality to that of films deposited with alternative techniques. Nowadays, PLD has become the deposition method to fabricate and study, in a research laboratory environment, epitaxial and single crystalline thin films of complex oxides, such as superconducting high-T_c cuprates, metaloxides, ferroelectric, ferromagnetic, and dielectric oxides. The success of PLD is due to the fact that it is conceptually as well as operationally a relatively simple technique. The main advantage of PLD is the possibility to transfer stoichiometrically multicomponent target material, especially containing volatile components such as lead, to the layer. Moreover, PLD is a powerful method which allows a thicker film to be fabricated in a short time due to the high deposition rate and the possibility to incorporate the process directly into a Si-production line [3, 4], for example in MEMS. Despite the fact that PLD technique has provided an excellent tool for producing high-quality multicomponent films, it has not yet fully emerged as a reproducible process for commercial thin-film applications. This is mainly because the deposited films have only a small area of structural and

IOP Conf. Series: Journal of Physics: Conf. Series **922** (2017) 012022 doi:10.1088/1742-6596/922/1/012022

thickness uniformity (samples of 10×10 mm^2 or smaller). Especially for the fabrication of MEMS, it is necessary to control the growth of these layers on silicon wafer-scales and to develop appropriate fabrication processes to realize micromachined structures [5, 6].

Microelectromechanical systems (MEMS) actuators driven by lead zirconate titanate at the morphotropic phase boundary, $Pb(Zr_{0.52}Ti_{0.48})O_3$ (abbreviated as PZT), have received wide attention recently because they could potentially outperform other MEMS actuators due to the remarkable high ferroelectric polarization and piezoelectric coefficients of PZT [7-11]. For the integration of PZT-based materials into MEMS these should be prepared in thin film form. Therefore the properties of PZT thin films are critical to the quality and reliability of MEMS devices. For most MEMS device applications, such as vibration energy harvesters, micropumps, microcantilever-based mass sensors and ultrasonic transducers for medical and sonar applications, the transverse piezoelectric coefficient (e_{31} or d_{31}) is the most important factor to be considered. For specific applications, however, such as a nano-control system for the control of optical cavities, a high longitudinal piezoelectric coefficient (d_{33}) is required to obtain a large piezoelectric, piston-like deformation in the PZT film actuators.

The subject of the present study is the deposition of $Pb(Zr_{0.52}Ti_{0.48})O_3$ (PZT) thin films on 4-inch platinized silicon wafers ($Pt/Ti/SiO_2/Si$ or short Pt/Si) using a large-area PLD tool developed by Solmates B.V. (a spin-off of the University of Twente–The Netherlands, was founded in 2007 and delivered the first large-area production tool for deposition of complex-oxide thin films on 4-8 inch wafers in 2013) [12]. The thickness homogeneity and microstructure uniformity of 1-μm-thick PZT thin films across wafers have been investigated. The piezoelectric MEMS devices, such as micro-cantilevers and micro-diaphragms were also fabricated. In addition, the high and uniformity of ferroelectric and piezoelectric properties of devices in the entire wafers indicated that the high-quality of PLD-based PZT films on Si wafer-scales will enhance the practical applicability of the piezoelectric MEMS devices.

2. Experimental procedure

Figure 1 shows a photograph of a Solmates's large-area PLD system in MESA+ NanoLab's cleanroom (University of Twente, The Netherlands). This system can handle wafers up to 200 mm in diameter and is specifically designed for PZT deposition, as well as the other perovskites and oxide materials, for MEMS applications. It includes wafer heating up to 800 °C, particle reduction filter, multi-target manipulator, fully automated with SEMI compatible fab interface software, integrated laser with homogenizer and beam attenuator. The fast growth rate of 3.4 μm/hour can be achieved for the deposition of PZT films.

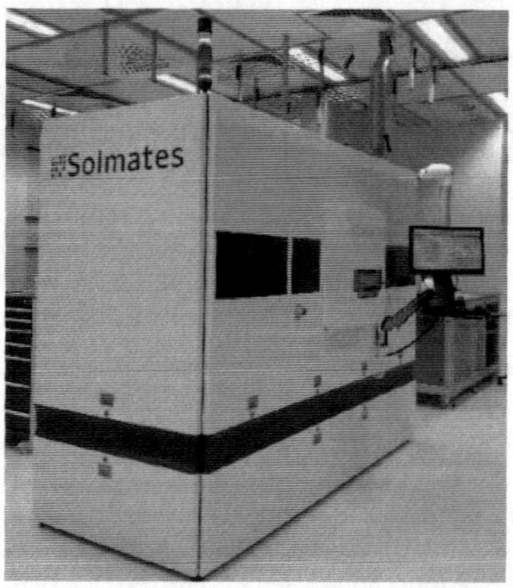

Figure 1. Solmates's Pulsed Laser thin-film Deposition (PLD) system in MESA+ NanoLab's cleanroom (University of Twente, The Netherlands).

In this study, the PZT films were deposited on 4-inch (111)$Pt/Ti/SiO_2/Si$ (Pt/Si) and 4-inch (111)$Pt/Ti/SiO_2/SOI$ (Pt/SOI, Silicon-on-Insulator) wafers using large-area PLD system with a KrF excimer laser. A 125-nm-thick Pt bottom electrodes and 15-nm-thick Ti adhesive layers were deposited at room temperature by DC magnetron sputtering on a 500-nm-thick SiO_2 layer formed through wet oxidation at 1100 °C. To prevent the formation of pyrochlore phases at the interface between the PZT film and the Pt bottom electrode, a thin nucleation layer of $LaNiO_3$ (LNO, 15 nm, using PLD) was inserted between these layers. The optimized growth conditions for the PZT and LNO layers were: substrate temperature of 600 °C; target-substrate distance 60 mm; laser spot size 6.2 mm^2; laser power density 3.0 J/cm^2 and a pressure of 0.1 mbar O_2 during deposition. After deposition, the films were cooled down to room temperature in a 0.1 mbar O_2 atmosphere. Top electrodes (Pt, 125 nm) were deposited on PZT films at room temperature by DC magnetron sputtering.

Crystallographic properties of the PZT films were analyzed with x-ray θ-2θ scans (XRD) using a PANalytical X'Pert X-ray Diffraction system. Film surface and microstructure was investigated using atomic force microscopy (AFM, Bruker Dimension ICON) and high-resolution scanning electron microscopy (HRSEM, Zeiss 1550). The Filmetrics F20 was used to measure the film thickness across the wafer.

For the electrical measurements, 300×300 μm^2 capacitors were patterned with a standard photolithography process and structured by argon-beam etching of the top Pt electrodes and wet-chemical etching (*HF-HCl* solution) of the PZT films to expose the bottom electrodes.

The polarization hysteresis (*P–E*) loop measurements were performed with the ferroelectric mode of the aixACCT TF-2000 Analyzer using a triangular *AC*-electric field of ±200 kV/cm at 1 kHz scanning frequency. All measurements were performed after bipolar cycling the devices a few times. The hysteresis loops did not change after the first few cycles. The longitudinal piezoelectric coefficient (d_{33f}) of the piezoelectric film capacitors was measured with a Double Beam Laser Interferometer (aixDBLI) apparatus that eliminates the influence of the substrate bending[13, 14], using a lock-in technique with a *DC* driving electrical field in the range ±200 kV/cm, and an *AC* peak-peak field amplitude of 5 kV cm^{-1} and 1 kHz frequency. The resolution of the aixDBLI is better than 1 pm/V. In addition, we measured the tip-displacement of 400×100×10 in μm PZT/Si cantilevers to determine the transverse piezoelectric coefficient (e_{31f}). The micro-cantilevers were produced from the PZT films grown on Pt/SOI wafer, and the tip deflection was measured with scanning laser Doppler vibrometer (LDV), driven by the piezoelectric stack at an *AC*-amplitude of ±30 kV/cm (at a *DC* offset voltage of 30 kV/cm) and 8 kHz frequency. The process for fabricating piezoelectric driven Si cantilevers has been described in a previous paper [15]. All measurements were performed at room temperature.

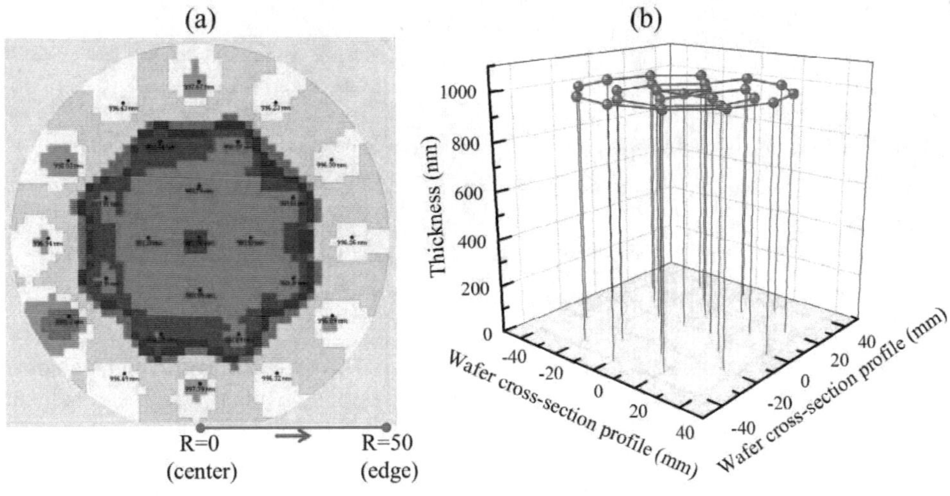

Figure 2. (a) A wafer map of 1-μm-thick PZT film deposited on 4-inch Pt/Si wafer produced by the Filmetrics system and (b) 3D-plot of the thickness uniformity.

IOP Conf. Series: Journal of Physics: Conf. Series **922** (2017) 012022 doi:10.1088/1742-6596/922/1/012022

3. Results and discussion

The film-thickness uniformity of PZT films on 4-inch Pt/Si wafers was measured using Filmetrics system. Figure 2 shows that the 1-µm-thick PZT films have been formed on 4-inch wafers with thickness variation of less than ±1% (990 nm ± 0.8%) across the wafers. The thickness variation also depends on the film thickness and it slightly increases with thicker films, for instance the 0.5 and 2-µm-thick PZT films have the thickness variation of about 0.4 and 2.2%, respectively.

In Figure 3, the crystallinity of PZT film deposited on 4-inch Pt/Si wafer is given. The XRD patterns at different wafer positions indicate that the relatively good uniformity of crystallinity and the preferred (001) orientation with minor (110) orientation are obtained in the film.

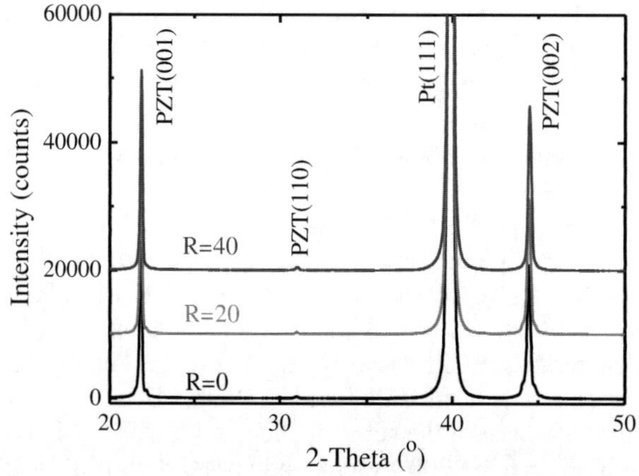

Figure 3. XRD patterns of PZT films at different wafer positions.

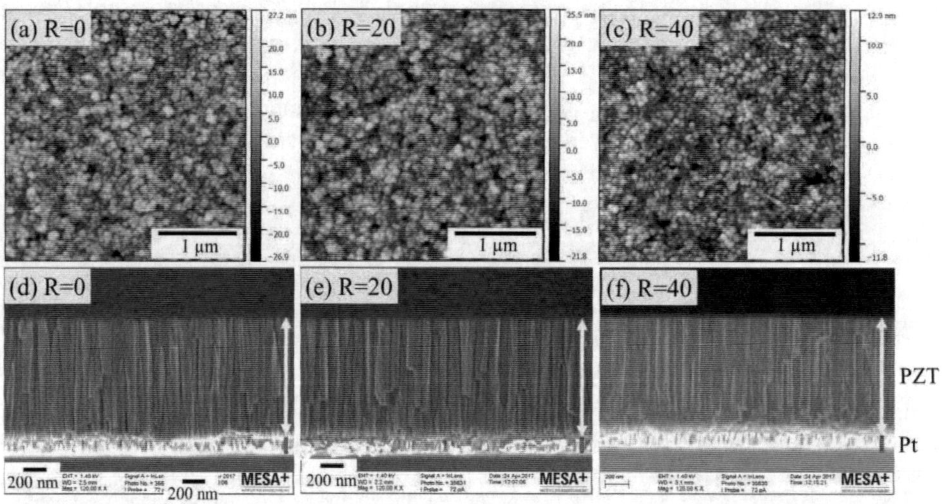

Figure 4. (a-c) Atomic force microscopy (AFM) and (d-f) cross-sectional scanning electron microscopy (SEM) of PZT film deposition on 4-inch Pt/Si wafer, at R=0 mm (wafer center), R=20 mm and R=40 mm (near wafer edge).

The surface morphology and microstructure of PZT films were investigated using atomic force microscopy (AFM) and cross-sectional scanning electron microscopy (SEM) measurements, respectively, as indicated in Figure 4. The columnar grains of PZT films are appeared from the Pt bottom electrodes and the grain size is increases with increasing film thickness. At the topper layer, the grain sizes in the range of 65-75 nm are obtained at the different wafer positions. The root-mean-square surface roughness (R_{rms}) value, obtained from AFM measurements, is slightly changed from the film at the wafer center (R_{rms} = 7.2 nm) to the film near the edge of wafer (R_{rms} = 4.0 nm). The details are also listed in Table 1. The change in the crystallinity and microstructure of PZT films is probably due to the laser scanning and/or the uniformity of wafer temperature during the deposition.

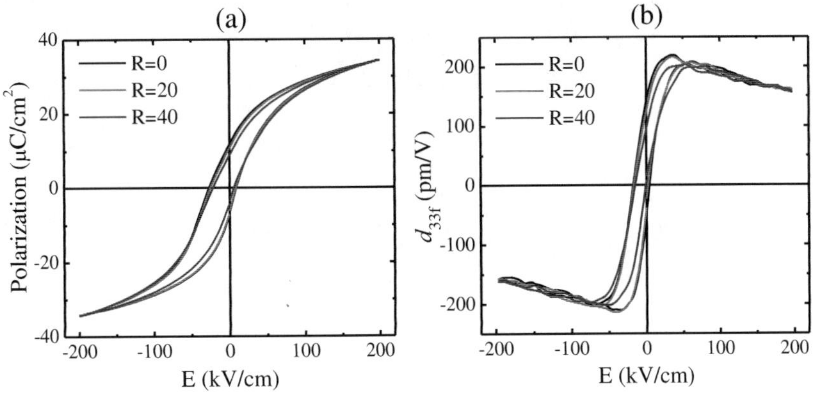

Figure 5. (a) Polarization (*P-E*) and (b) piezoelectric (d_{33f}-*E*) hysteresis loops of PZT films, measured at different wafer positions.

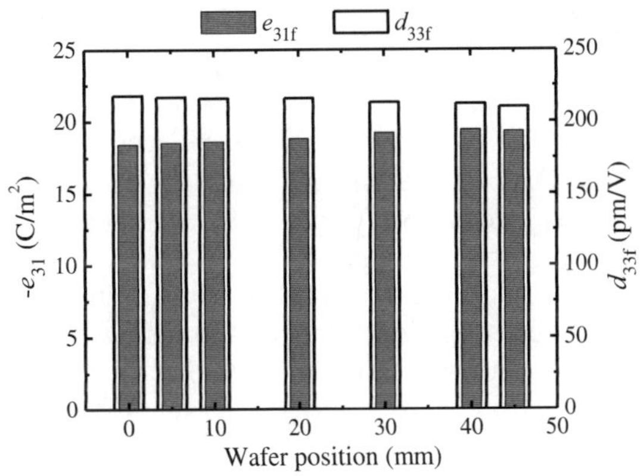

Figure 6. Relationship between piezoelectric e_{31f} and d_{33f} coefficients and position of 4-inch PZT/SOI wafer.

In order to investigate the ferroelectric and piezoelectric properties of PZT films, the polarization (*P-E*) and piezoelectric (d_{33f}-*E*) hysteresis loops were performed at ±200 kV/cm and 1 kHz, as shown in Figure 5. The results indicate that the remanent polarization (P_r), coercive field (E_c) and longitudinal piezoelectric coefficient (d_{33f}), defined from the PZT film capacitors (300×300 μm^2 in size), are higher at the wafer center. Whereas, the transverse piezoelectric coefficient (e_{31f}) value is slightly enhanced in the devices near the wafer edge (Figure 6). In this case, the more (001) orientation or less (110) orientation in PZT film at the wafer center has a larger P_r value because the spontaneous polarization is along the [001] crystallographic direction. Meanwhile, the enhanced d_{33f} value at wafer center is due to the less closely packed columnar grains, which was previously ascribed to increased declamping of the columns in the film and therefore easier domain wall motion and an increased extrinsic piezoelectric effect [16]. However, the less compacted columnar grain is a possible reason for the lower e_{31f} coefficient from the devices at the wafer center [17].

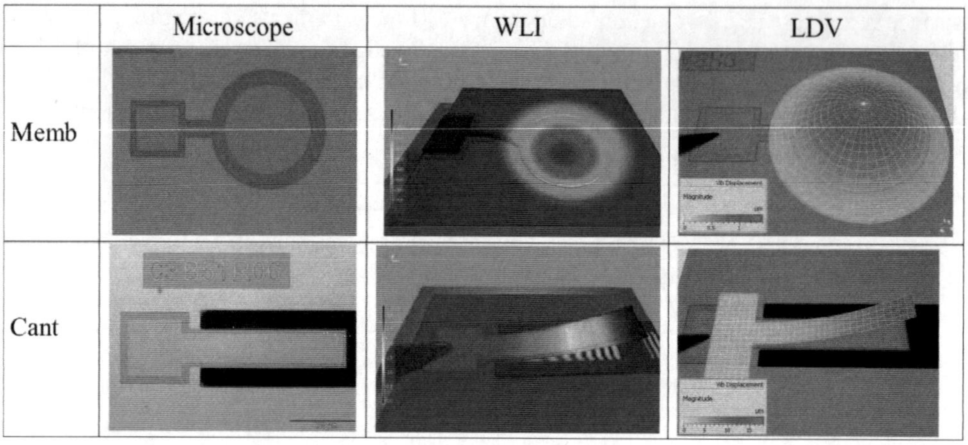

Figure 7. Optical microscope, white-light interferometer (WLI) initial bending and laser Doppler vibrometer (LDV) piezoelectric displacement of a piezoelectric micro-membrane (Memb: 500-μm diameter) and a micro-cantilever (Cant: 400-μm×100-μm, length×width) based on 1-μm-thick PZT films on Pt/10-μm-thick Si device layers.

As mentioned above, there are many potential applications of piezoelectric thin films in the MEMS area. Through large-are PLD technique and wafer-scale microfabrication, the devices based on piezoelectric thin films with different structures can be simultaneously fabricated. Two of these structures, membrane micropumps and resonant micro-cantilever mass-sensors, are shown in Figure 7. The displacement at the center of 500-μm-diameter membrane is about 1.4 μm/V (at the first-mode resonant-frequency of 464 kHz). The quality factor and mass-sensitivity are about 500 and 1 pg/Hz, respectively, for the micro-cantilevers (400-μm×100-μm, length×width) defined at the first-mode resonant peak (with the resonant frequency of 72 kHz). The mass-sensitivity in the cantilevers based PLD films this study is much better than those using the sol-gel (mass-sensitivity: 30.3 pg/Hz) [18] or screen-printing (mass-sensitivity: 62.5 pg/Hz) [19] methods with similar cantilever structure. Moreover, the high yield wafer-scale fabrications of MEMS-based PLD PZT/Si-devices have also obtained. The high uniformity in MEMS devices in the wafer can be seen from the stability in the transverse e_{31f} coefficients where defined from the displacement of cantilevers at different wafer positions (Figure 6).

Table 1. Properties of 1-μm-thick PZT films on 4-inch Pt/Si at different wafer positions.

	Wafer position		
	R=0 mm (center)	R=20 mm	R=40 mm (near edge)
R_{rms} (nm)	7.2	6.5	4.0
P_r (μC/cm^2)	11.8	11.0	9.2
E_c (kV/cm)	18.2	17.1	14.0
d_{33f} (pm/V)	218	216	210
e_{31f} (C/m^2)	-18.4	-18.8	-19.3

4. Conclusion

Piezoelectric PZT thin films have been in-situ deposited on 4-inch Pt/Si or Pt/SOI wafers with large-area pulsed laser deposition (PLD) technique. A high deposition rate of 55 nm/min is obtained. Pure perovskite phase with preferred (001) orientation of the PZT films is achieved. High uniformity in PZT films can be observed from the thickness, microstructure, ferroelectric and piezoelectric properties. The high device yield and reproducibility of high-performance MEMS-based *PLD* PZT/wafer-scales will enhance the applicability of piezoMEMS devices in the industrial applications.

References

[1] H. M. Smith and A. F. Turner, "Vacuum deposited thin films using a ruby laser," *Appl. Opt.* 4 (1965) 147-148.

[2] D. Dijkkamp, T. Venkatesan, X. D. Wu, S. A. Shaheen, N. Jisrawi, Y. H. Min-Lee, W. L. McLean, and M. Croft, "Preparation of Y-Ba-Cu oxide superconductor thin films using pulsed laser evaporation from high T_c bulk material," *Appl. Phys. Lett.* 51 (1987) 619-621.

[3] M. Jelínek, V. Trtík, and L. Jastrabík, "Pulsed laser deposition of thin films," in *Physics and Materials Science of High Temperature Superconductors, IV*, Editors: M.J. R. Kossowsky and J. Novak, Springer: Dordrecht-Netherlands, 1997, pp. 215-231.

[4] M. D. Nguyen, H. N. Vu, D. H. A. Blank, G. Rijnders, "Epitaxial $Pb(Zr,Ti)O_3$ thin films for a MEMS application," *Adv. Nat. Sci.: Nanosci. Nanotechnol.* 2 (2011) 015005.

[5] J. F. Scott, "Applications of modern ferroelectrics," *Science* 315 (2007) 954-959.

[6] Y. Wang, C. Ganpule, B. T. Liu, H. Li, K. Mori, B. Hill, M. Wuttig, R. Ramesh, J. Finder, Z. Yu, R. Droopad, and K. Eisenbeiser, "Epitaxial ferroelectric $Pb(Zr,Ti)O_3$ thin films on Si using $SrTiO_3$ template layers," *Appl. Phys. Lett.* 80 (2002) 97-99.

[7] C. B. Eom and S. Trolier-McKinstry, "Thin-film piezoelectric MEMS," *MRS Bulletin* 37 (2012) 1007-1017.

[8] G. L. Smith, J. S. Pulskamp, L. M. Sanchez, D. M. Potrepka, R. M. Proie, T. G. Ivanov, R. Q. Rudy, W. D. Nothwang, S. S. Bedair, C. D. Meyer, and R. G. Polcawich, "PZT-based piezoelectric MEMS technology," *J. Am. Ceram. Soc.* 95 (2012) 1777-1792.

[9] T. Iijima, S. Osone, Y. Shimojo, and H. Nagai, "Synthesis of 10-μm-thick lead zirconate titanate films on 2-in. Si substrates for piezoelectric film devices," *Int. J. Appl. Ceram. Technol.* 3 (2006) 442-447.

[10] K. Kanda, I. Kanno, H. Kotera, and K. Wasa, "Simple fabrication of metal-based piezoelectric MEMS by direct deposition of $Pb(Zr,Ti)O_3$ thin films on titanium substrates," *J. Microelectromech. Syst.* 18 (2009) 610-615.

[11] P. Muralt, "Recent progress in materials issues for piezoelectric MEMS," *J. Am. Ceram. Soc.* 91 (2008) 1385-1396.

[12] http://www.solmates.nl/.

[13] K. Prume, P. Muralt, F. Calame, T. Schmitz-Kempen, and S. Tiedke, "Extensive electromechanical characterization of PZT thin films for MEMS applications by electrical and mechanical excitation signals," *J. Electroceram.* 19 (2007) 407-411.

[14] K. Prume, P. Muralt, F. Calame, T. Schmitz-Kempen, and S. Tiedke, "Piezoelectric thin films: evaluation of electrical and electromechanical characteristics for MEMS devices," *IEEE Trans. Ultras. Ferroelectr. Freq. Contr.* 10 (2005) 8-14.

[15] M. D. Nguyen, M. Dekkers, H. N. Vu, and G. Rijnders, "Film-thickness and composition dependence of epitaxial thin-film PZT-based mass-sensors," *Sens. Act. A: Phys.* 199 (2013) 98-105.

[16] M. D. Nguyen, E. P. Houwman, M. Dekkers, and G. Rijnders, "Strongly enhanced piezoelectric response in lead zirconate titanate films with vertically aligned columnar grains," *ACS Appl. Mater. Interfaces* 9 (2017) 9849-9861.

[17] M. D. Nguyen, H. Yuan, E. P. Houwman, M. Dekkers, G. Koster, J. E. ten Elshof, and G. Rijnders, "Highly oriented growth of piezoelectric thin films on silicon using two-dimensional nanosheets as growth template layer," *ACS Appl. Mater. Interfaces* 8 (2016) 31120–31127.

[18] S. Shin, S. Song, Y. Lee, N. Lee, J. Park, H. Park, and J. Lee, "Fabrication and sensing behavior of piezoelectric microcantilever for nanobalance," *Jpn. J. Appl. Phys.* 42 (2003) 6139-6142.

[19] J. H. Park, H. Kim, D. S. Yoon, S. Y. Kwang, J. Lee, and T. S. Kim, "Effects of the material properties on piezoelectric PZT thick film micro cantilevers as sensors and self actuators," *J. Electroceram.* 25 (2010) 1-10.

IOP Conf. Series: Journal of Physics: Conf. Series **922** (2017) 012023 doi:10.1088/1742-6596/922/1/012023

Enhanced gas sensing performance of graphene/ZnS-CdS hetero-nanowires gas sensor synthesized by Langmuir-Blodgett self-assembly method

Hu Li[1], Ishtiaq Hassan Wani[1], Anumol Ashok[1], Yuanyuan Han[1], S. Hassan M. Jafri[1], Somobrata Acharya[2], Klaus Leifer[1]

[1] Ångström Laboratory, Uppsala University, Sweden.

[2] Centre for Advanced Materials (CAM), Indian Association for the Cultivation of Science, India

Klaus.Leifer@angstrom.uu.se

Abstract.

Graphene is a promising material in the field of solid-state gas sensors due to the unique two-dimensional structure. Here, we have shown by fabricating graphene/ZnS-CdS hetero-nanowire structure, the gas sensor sensitivity has a two-fold increase to 20% under 15 ppm gaseous concentration compared to a 10% response in pristine graphene. Spectroscopy and microscopy analysis indicate that the semi-conducting ZnS-CdS hetero-nanowires are 2 nm wide and densely packed on top of graphene. By combining UV illumination, the device approaches a fast response/recovery and high gas sensitivity, thus has a potential to be used in a detection of wide range of gases.

Introduction

Due to the larger surface/volume ratio and low electrical noises, graphene has shown outstanding gas sensing properties [1–4] and even the single molecule detection [5]. Through modifying the carrier density of graphene, the absorbed gaseous molecules perform as electron acceptors or donors and hence change the electrical conductivity of the graphene [4]. Various p- and n-doped gas molecules have been systematically studies to understand the physical sensing mechanism of graphene and it is found that p- doped gases result in an increase in the conductance of the graphene, while the n-doped gases lead to a decrease in the conductance [5]. Among the gas molecules, NO_2 has a rather high charge transfer and exhibits good sensitivity to pristine graphene, thus is commonly used in the electrical gas sensing measurements.

Semi-conductor nanowires have attracted considerable concerns owing to their potential applications ranging from electronics to biology [6–8]. Due to the large surface area and the semi-conductor property, they have shown excellent gas sensing performance in a variety of gases [9,10]. In this work, through the bottom up approach, the gas sensing devices were fabricated by physically attaching well aligned ZnS-CdS nanowires on graphene surface and the electrical gas sensitivity measurements were performed to investigate the modification of the sensitivity. The result indicates that, compared with the response of 10% in pristine graphene, the electrical response of the fabricated devices have shown a two –fold increase to ~20% under the gaseous concentration of 15 ppm. Meanwhile, by employing the UV light improved desorption, the devices show a fast recovery to the initial states. Therefore, the demonstrated gas sensor has the potential to be used in a wide range of gas sensing applications.

Experimental Details

2.1 Materials

In this study, large scale graphene synthesized by chemical vapor deposition (CVD) method were used and transferred to a SiO_2/Si substrate [11]. The ZdS-CdS hetero-nanowires were synthesized by using a wet-chemical technique in organic medium [12] and the transfer of the ZdS-CdS hetero-nanowires onto graphene surface was done by using a Langmuir–Blodgett film technique [13].

2.2 Device fabrications

Electrical contacts on graphene, as shown in Figure 1a, were fabricated by using a photolithography technique [14], and 10 nm Ti and 150 nm Au were deposited using evaporation followed by a metal lift-off process in acetone. The device after the transfer of the Langmuir-Blodgett self-assembled ZnS-CdS hetero-nanowires is shown in Figure 1b.

2.3 Characterizations

The optical observation, atomic force microscopy, Raman spectroscopy and light absorption spectroscopy were performed in a Olympus light optical microscope, a Bruker Multimode 8 AFM, a Renishaw inVia Raman spectroscopy with 532 nm excitation and a Shimadzu UV-VIS Spectrometer, respectively. The photoconductivity measurement was done by using a DH-2000 Ocean UV-VIS-NIR light source and a Keithley 6430 sub-femtoamp remote sourcemeter. The electrical sensitivity characterization was carried out in a Keithley 6430 sub-femtoamp source-meter in a gas chamber and a mixture of N_2 and 15 ppm NO_2 was chosen the target gas in the measurement.

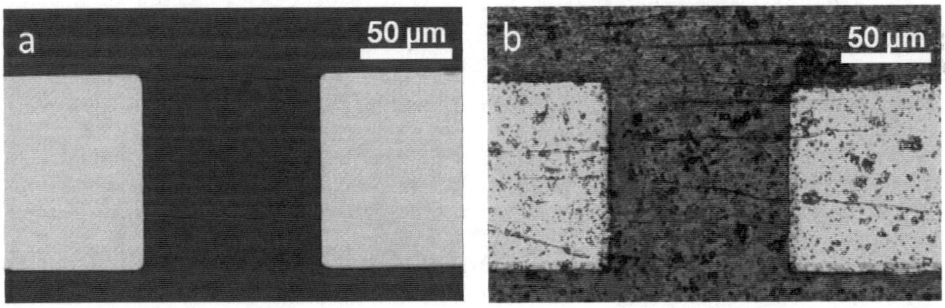

Figure 1. Light optical microscopy images of (a) graphene with electrical contact, (b) after the ZnS-CdS hetero-nanowires transfer.

Result and Discussion

Before and after the ZnS-CdS hetero-nanowires deposition, Raman spectroscopy was carried out to investigate the possible structure modification of the graphene, as shown in Figure 2. It can be seen that both of the spectra consist in two main peaks: 2D peak at at 2700 cm^{-1} and G peak at at 1580 cm^{-1}. The sharp 2D peak combing the low intensity of the D peak (at 1350 cm^{-1}, that shows the defect density) indicates that the ZnS-CdS nanowires are physically absorbed on graphene surface and no chemical bond forms, thus the ZnS-CdS hetero-nanowires functionalized graphene should maintain the superior properties of the graphene.

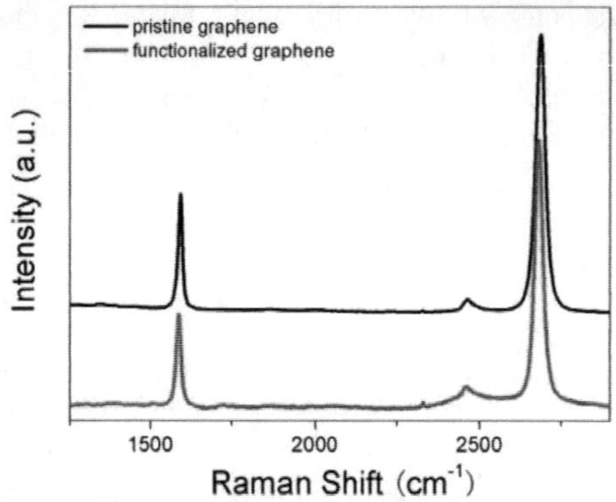

Figure 2. Raman comparison of the pristine graphene (black) and functionalized graphene (red).

The ZnS-CdS hetero-nanowires were prepared by a wet chemical approach [12], where the structure sketch is shown in the inserted image of Figure 3. The hetero-nanowires consist of periodic constituent blocks of semiconductor ZnS and CdS. The absorption property in solution status was investigated and shown in Figure 3. UV-VIS absorption spectrum exhibits that the absorption onsets are at 340 nm and 440nm, corresponding to the ZnS bandgap of 3.65 eV and CdS bandgap of 2.82 eV, respectively. The nanowires then were transferred onto graphene surface by using a Langmuir–Blodgett film technique [13]. The surface topography [15–17] was studied by using AFM and the high resolution height and peak force error images are shown in Figure 4. It can be seen that the ZnS-CdS hetero-nanowires after transfer are ~2 nm in diameter and were densely packed on the surface.

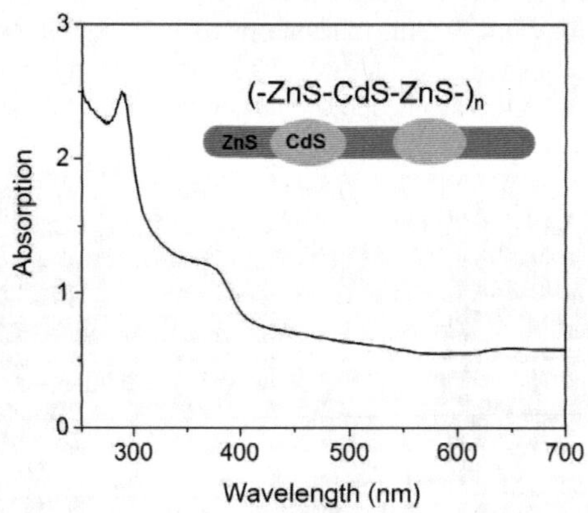

Figure 3. UV absoption spectrum of the ZnS-CdS hetero-nanowires

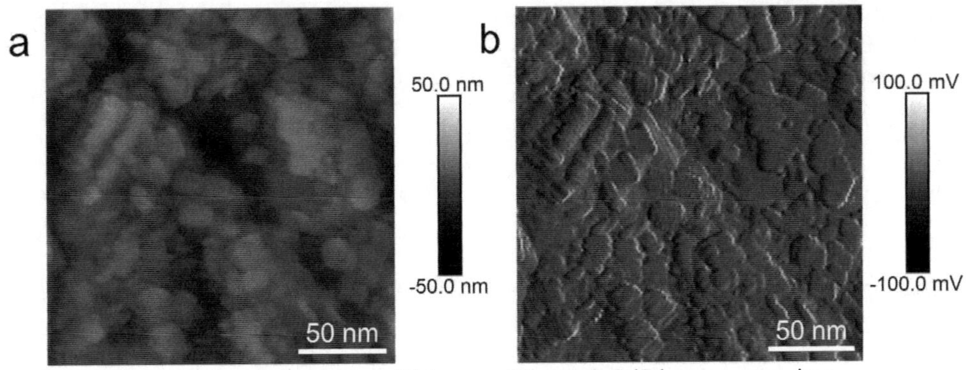

Figure 4. AFM height (a) peak force error (b) image of the ZnS-CdS hetero-nanowires

Due to the semi-conductor property of the ZnS-CdS hetero-nanowires, under UV light illumination, photoelectron can be generated and contribute to the current, which is confirmed in the photoconductivity measurement shown in Figure 5. In this experiment, under the bias voltage of 1 mV, the initial current of the functionalized graphene was 1.08 μA and increases to 1.13μA under illumination.

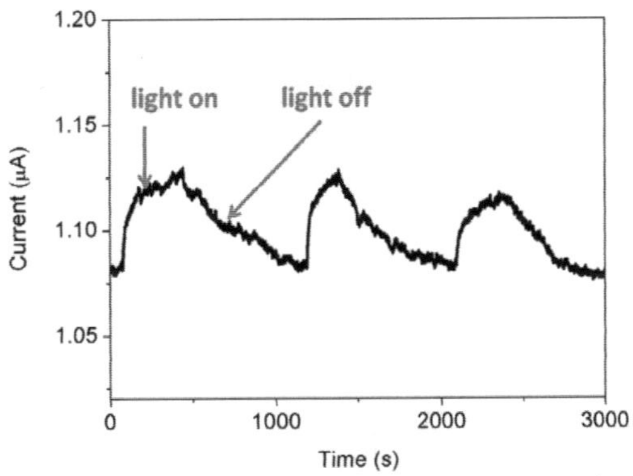

Figure 5. Photoconductivity measurement of the ZnS-CdS hetero-nanowires functionalized graphene under the bias voltage of 1mV.

Figure 6 shows the normalized conductance response of the pristine graphene and functionalized graphene under the exposure of 15 ppm NO_2 in N_2. The electrical response magnitude S was expressed as:

$$S = \left(\frac{G}{G_0} - 1\right) \times 100\%$$

where G_0 is the measured conductance of graphene prior to the NO_2 exposure, i.e. when exposing to pure N_2 gas and G is the measured conductance of graphene under the NO_2 exposure. When the pristine graphene is exposed to NO_2 gas, the electrical conductance of the graphene has a significant increase of 10% and this is because of the electron acceptor role of NO_2 in the sensing, where the electrons are transferred to the NO_2 molecules from graphene and finally leading to an increase in the hole density of graphene. The response time in graphene gas sensing was found to be ~1000s and, to accelerate the recovery process, the UV light illumination was employed to desorb the NO_2 molecules to initialize the device [4]. As for the ZnS-CdS hetero-nanowires functionalized graphene, the sensitivity has a dramatic increase to 20%, which is two-fold as high as pristine graphene, and the response time is decreased to ~300s, which is one third as compared to pristine graphene. Therefore, the demonstrated gas sensor has considerably enhanced the response time as well as the sensitivity of pristine graphene.

This improved sensing performance of the ZnS-CdS hetero-nanowires functionalized graphene can be explained from the perspective of absorption energy and surface sites. According to the work of Tamvakos et al [10], the NO_2 is tended to bond to Zn atoms due to a high absorption energy. Thus, the faster response in the functionalized graphene can be well explained considering the comparably low absorption energy between NO_2 molecules and pristine graphene. Moreover, the large surface area of the ZnS-CdS hetero-nanowires provide a large amount of sites for the NO_2 molecules to position, combining with the higher absorption energy, thus the sensitivity is dramatically enhanced as well.

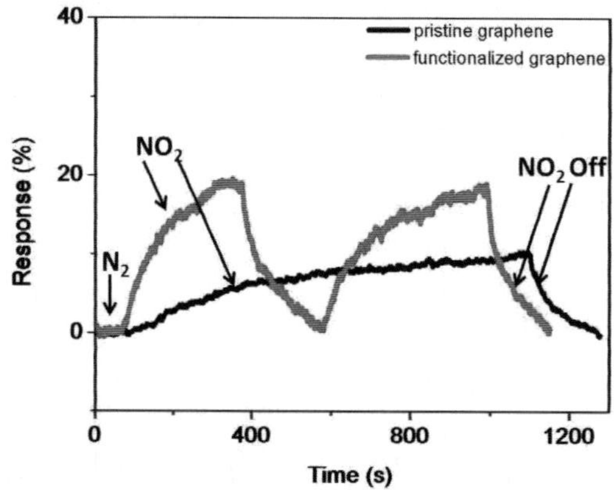

Figure 6. Normalized conductive response of graphene gas sensors.

Conclusion

In our work, the ZnS-CdS hetero-nanowires functionalized graphene were successfully synthesized and incorporated into the gas sensor device. The semiconducting ZnS-CdS hetero-nanowires show the absorption onsets of 340 nm and 440 nm corresponding to the band gap of 3.65 eV for ZnS and 2.82 eV for CdS, respectively. Raman spectroscopy and AFM analysis indicates that the 2 nm wide nanowires are physically arranged on graphene surface and there is no chemical bond formed. Through the electrical gas sensing measurements, it is found that the hybrid material shows much faster response and the normalized conductive response of the ZnS-CdS hetero-nanowires functionalized graphene has a two-fold increase to 20% compared with the response of pristine graphene of 10%. Combing the UV illumination to accelerate the molecule desorption process, the devices exhibit fast response and recovery and, thus, has the possibility to be used in application of wide range of gases detection .

Acknowledgements

We gratefully acknowledge the financial support from Knut and Alice Wallenberg Foundation and Swedish Research Council (Vetenskapsrådet).

References

[1] Geim a K and Novoselov K S 2007 The rise of graphene *Nat. Mater.* **6** 183–91
[2] Li H, Daukiya L, Haldar S, Lindblad A, Sanyal B, Eriksson O, Aubel D, Hajjar-Garreau S, Simon L and Leifer K 2016 Site-selective local fluorination of graphene induced by focused ion beam irradiation *Sci. Rep.* **6** 19719
[3] Papadakis R, Li H, Bergman J, Lundstedt A, Jorner K, Ayub R, Haldar S, Jahn B O, Denisova A, Zietz B, Lindh R, Sanyal B, Grennberg H, Leifer K and Ottosson H 2016 Metal-free

photochemical silylations and transfer hydrogenations of benzenoid hydrocarbons and graphene *Nat. Commun.* **7** 12962

[4] Hajati Y, Blom T, Jafri S H M, Haldar S, Bhandary S, Shoushtari M Z, Eriksson O, Sanyal B and Leifer K 2012 Improved gas sensing activity in structurally defected bilayer graphene. *Nanotechnology* **23** 505501

[5] Schedin F, Geim A, Morozov S, Hill E, Blake P, Katsnelson M and Novoselov K 2007 Detection of individual gas molecules adsorbed on graphene *Nat. Mater.* **6** 652–5

[6] Yang P, Yan R and Fardy M 2010 Semiconductor nanowire: Whats Next? *Nano Lett.* **10** 1529–36

[7] Peidong Yang 2010 Semiconductor nanowires for energy conversion *2010 3rd Int. Nanoelectron. Conf.* 49–49

[8] Rahong S, Yasui T, Kaji N and Baba Y 2016 Recent developments in nanowires for bio-applications from molecular to cellular levels *Lab Chip* **16** 1126–38

[9] Jońca J, Ryzhikov A, Kahn M L, Fajerwerg K, Chaudret B, Chapelle A, Menini P and Fau P 2014 Shape-controlled ZnO Nanostructures for Gas Sensing Applications *Procedia Eng.* **87** 907–10

[10] Tamvakos A, Korir K, Tamvakos D, Calestani D, Cicero G and Pullini D 2016 NO_2 Gas Sensing Mechanism of ZnO Thin-Film Transducers: Physical Experiment and Theoretical Correlation Study *ACS Sensors* **1** 406–12

[11] Liang X, Sperling B a, Calizo I, Cheng G, Hacker C A, Zhang Q, Obeng Y, Yan K, Peng H, Li Q, Zhu X, Yuan H, Walker A R H, Liu Z, Peng L-M and Richter C a 2011 Toward clean and crackless transfer of graphene. *ACS Nano* **5** 9144–53

[12] Pradhan N, Acharya S, Ariga K, Karan N S, Sarma D D, Wada Y, Efrima S and Golan Y 2010 Chemically programmed ultrahigh density two-dimensional semiconductor superlattice array *J. Am. Chem. Soc.* **132** 1212–3

[13] Acharya S, Hill J P and Ariga K 2009 Soft langmuir-blodgett technique for hard nanomaterials *Adv. Mater.* **21** 2959–81

[14] Li H, Wani I H, Hayat A, Jafri S H M and Leifer K 2015 Fabrication of reproducible sub-5 nm nanogaps by a focused ion beam and observation of Fowler-Nordheim tunneling *Appl. Phys. Lett.* **107** 103108

[15] Cai Y, Li H, Karlsson M, Leifer K, Engqvist H, Xia W, Plodinec M, Loparic M, Monnier C a, Obermann E C, Zanetti-Dallenbach R, Oertle P, Hyotyla J T, Aebi U, Bentires-Alj M, Lim R Y H and Schoenenberger C-A 2016 Biomineralization on single crystalline rutile: the modulated growth of hydroxyapatite by fibronectin in a simulated body fluid *RSC Adv.* **6** 35507–16

[16] Bhattacharya K, El-Sayed R, Andón F T, Mukherjee S P, Gregory J, Li H, Zhao Y, Seo W, Fornara A, Brandner B, Toprak M S, Leifer K, Star A and Fadeel B 2015 Lactoperoxidase-mediated degradation of single-walled carbon nanotubes in the presence of pulmonary surfactant *Carbon N. Y.* **91** 506–17

[17] El-Sayed R, Bhattacharya K, Gu Z, Yang Z, Weber J K, Li H, Leifer K, Zhao Y, Toprak M S, Zhou R and Fadeel B 2016 Single-Walled Carbon Nanotubes Inhibit the Cytochrome P450 Enzyme, CYP3A4. *Sci. Rep.* **6** 21316